U0829855

天津历史风貌建筑
HISTORIC ARCHITECTURE IN TIANJIN

公共建筑卷二
Public Buildings Volume 2

路红 主编

天津大学出版社
TIANJIN UNIVERSITY PRESS

《天津历史风貌建筑丛书》编写组
AUTHORS GROUP

主　　编：路　红

副 主 编：唐连蒙　徐连和　孙　超　关　静

编　　辑：王志青　张俊东　李琦琳　孔　晖　金彭育　陈广发　孙　巍
　　　　　李　莉　谈卫卫　肖　楠　姚　鑫　袁　勇　何效星

摄　　影：何　方　何　易

Editor-in-chief: Lu Hong

Associate Editors: Tang Lianmeng, Xu Lianhe, Sun Chao, Guan Jing

Editors: Wang Zhiqing, Zhang Jundong, Li Qilin, Kong Hui, Jin Pengyu, Chen Guangfa, Sun Wei, Li Li, Tan Weiwei, Xiao Nan, Yao Xin, Yuan Yong, He Xiaoxing

Photographers: He Fang, He Yi

目录
CONTENTS

Financial Buildings
金融建筑

Former Hongkong and Shanghai Bank Corporation

原汇丰银行

汇丰银行创办于1864年，总行设在香港。天津分行1881年开业，为天津第一家外商银行，是天津对外贸易的主要结算银行，获有中国关税、盐税的托管权，享有纸币发行权，决定着天津的外汇市场。1954年撤离。该建筑位于和平区解放北路82号，天津市文物保护单位，特殊保护等级历史风貌建筑，现为中国银行。建筑建于1925年，占地面积约2 900 m^2，建筑面积约4 300 m^2。由英商爱迪克生和达拉斯（同和）工程司设计，钢混结构三层楼房。建筑正面及侧面矗立十二根巨大的爱奥尼克石柱，体现出古罗马柱式的高大雄伟，属典型的古典主义风格。

Hongkong and Shanghai Bank Corporation (HSBC) was founded in 1864 with its headquarter in Hong Kong. HSBC Tianjin Branch opened in 1881, being the first foreign capital bank in the city. It served as the main international settlement & trustee bank, dealing with the foreign trade, custom duty & salt tax, and banknote issue in the city.

Before it ended its business in 1954, the Branch Bank was located at 82 North Jiefang Road, Heping District. Designed by the English designing body, Atkinson & Dallas Architects and Civil Engineers Ltd., the building constructed in 1925 has been listed as the heritage site under city protection and the special protection historic architecture in Tianjin. Now, the building is used by Bank of China, covering an area of about 2, 900 square meters, and about 4 ,300 square meters in floor area.

With 12 massive Ionic columns erecting along its facade and sides in the ancient Roman orders, the three-storied reinforced concrete structure looks grand and majestic. It is featured to be in a Classical style.

历史照片 Historic photo

主入口 Main Entrancev

外景 Exterior view

中国银行

外景营业厅内景 Inside Business Hall

马赛克铺装地面 Mosaic pavements

老电梯 Old-fashioned elevator

外景 Exterior view

Former Chartered Bank of India, Australia and China

原麦加利银行

麦加利银行（亦称渣打银行），创办于1853年，总行设于伦敦，是英国皇家特许的殖民地银行。天津分行开业于1895年，主要业务为定活期存款、汇兑信用证，1954年关闭。该建筑位于和平区解放北路149~153号，天津市文物保护单位，特殊保护等级历史风貌建筑，现为办公用房。建筑建于1926年，占地面积约2 300 m^2，建筑面积约5 600 m^2，由英商赫明与帕尔克因（景明）工程公司设计，二层钢混结构楼房（设有地下室）。建筑主入口由六棵爱奥尼克式巨柱形成开敞柱廊，气势宏伟庄严，建筑整体感很强，是典型的古典主义风格。

历史照片 Historic photo

Also known as the Standard Chartered Bank, the Chartered Bank of India, Australia and China was founded via the approval of the English Royalty in 1853 with its headquarter located in London. It could be identified by its name as one of the chartered banks specially operating in the English colonies. It opened its Tianjin Branch in 1895, with the business scope covering the current deposit and exchanging LC (Letter of Credit), etc. The Branch ends its business in 1954. The building formerly serving for the Branch is located at 149~153 North Jiefang Road, Heping District. It has been regarded as the heritage site under city protection and the special protection historic architecture in Tianjin. Now, the building is used as office building, covering an area of about 2,300 square meters, and about 5,600 square meters in floor area.

The three-storied building was designed by the Hemmings & Parkin Ltd., an English engineering company. The open colonnade with 6 massive Ionic columns is added to the Main Entrance of this reinforced concrete structure, looking more majestic and spectacular as a whole. It is a Classical architecture in style.

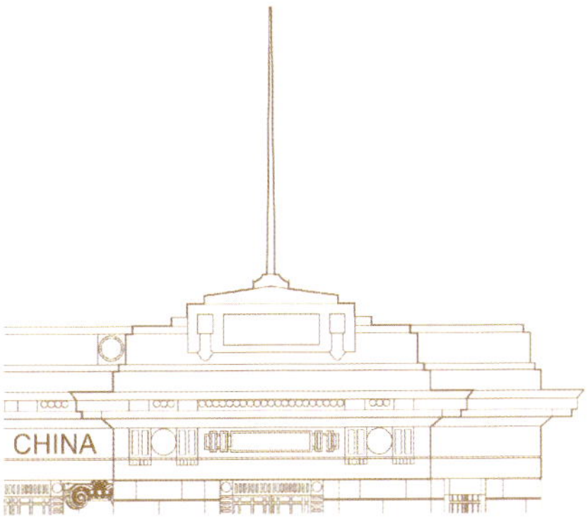

正立面渲染图 Facade Rendering

外景 Exterior view

檐口与柱式 Cornice and column order

侧入口 Side Entrance

外檐局部 Partial outer-eaves

外檐窗 Windows below the oute-eaves

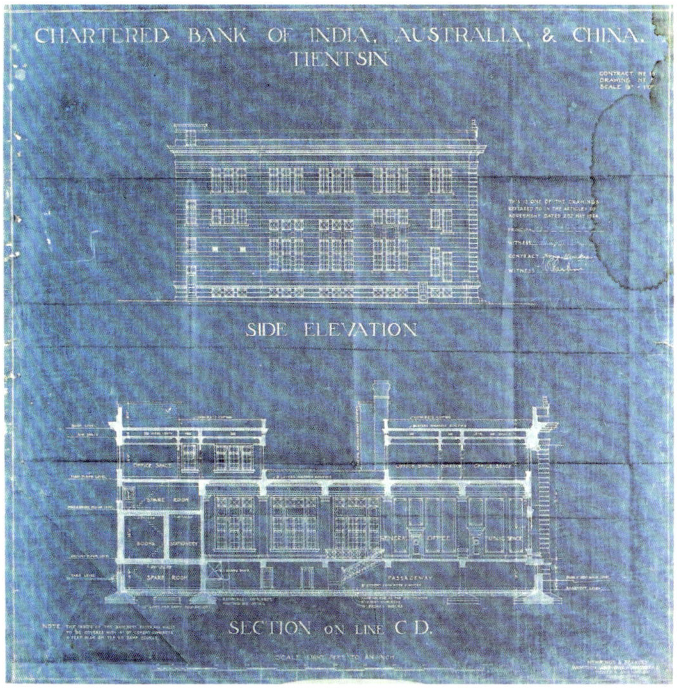

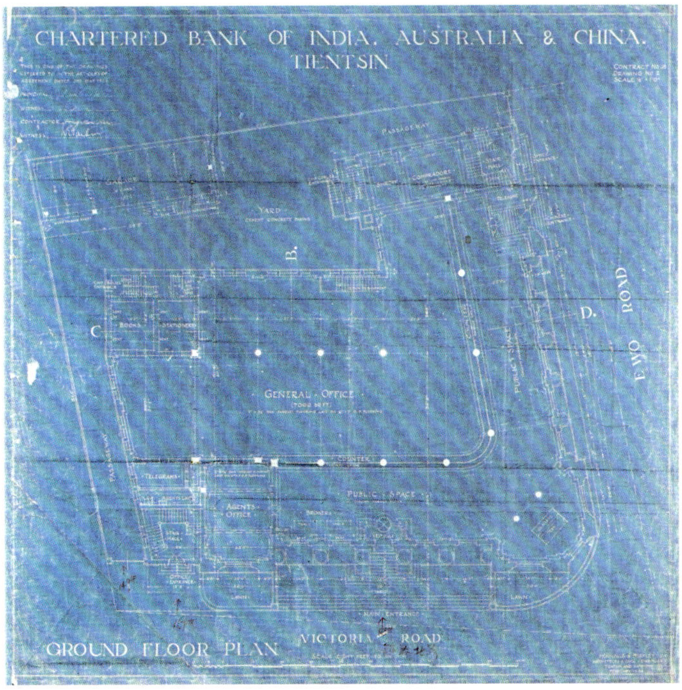

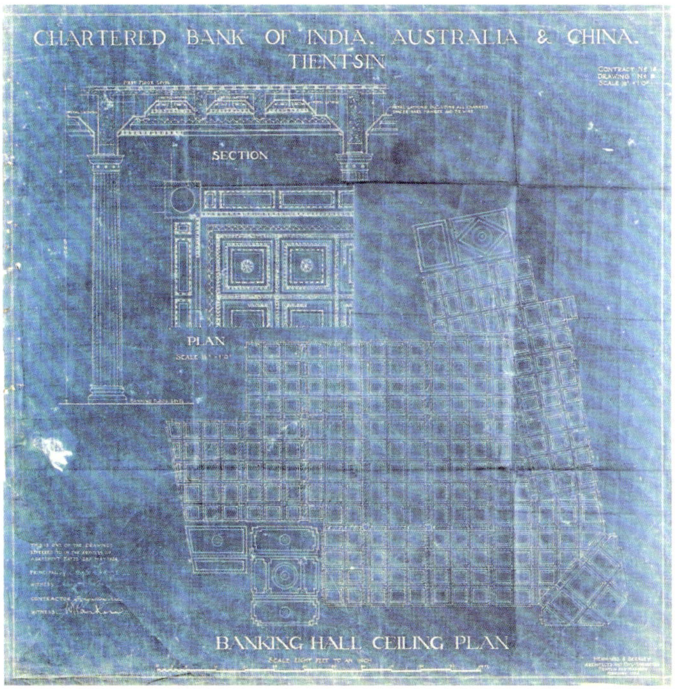

原设计图 Original blueprint

营业厅内门楼 Archway inside the Business Hall

营业厅顶棚 Ceiling of the Business Hall

营业厅内门楼细部
Archway inside the Business Hall

老暖气罩铜花饰
Copper decoration to the old-fashioned radiator enclosure

Former L'indo Bank of China

原东方汇理银行

原东方汇理银行位于和平区解放北路77~79号，天津市区、县文物保护单位，重点保护等级历史风貌建筑，现为办公用房。东方汇理银行创办于1875年，总行设于法国巴黎。二十世纪初先后在我国上海、天津、北京、汉口、广州等地建立分行。天津分行1907年开业，主要经营进、出口押汇及买卖外汇。1913年曾代表法国参与五国银行团，与北洋政府签订了“善后大借款”。太平洋战争爆发后，该行利用政治手段，继续经营，取得了外商银行在天津的首席地位。中华人民共和国成立后，经中国银行批准，该行一度成为办理外汇的指定银行，1957年停业。该建筑1912年建成，占地面积约1 200 m^2，建筑面积约3 500 m^2，比商义品公司设计。砖木结构三层楼房，建筑首层为水泥仿石块墙面，二、三层为清水红砖墙与混水墙面搭配，墙面装饰有抹灰纹样。建筑造型稳重，细部丰富，色彩明快大方，具有西洋古典主义建筑特征。

外景 Exterior view

西洋美术馆

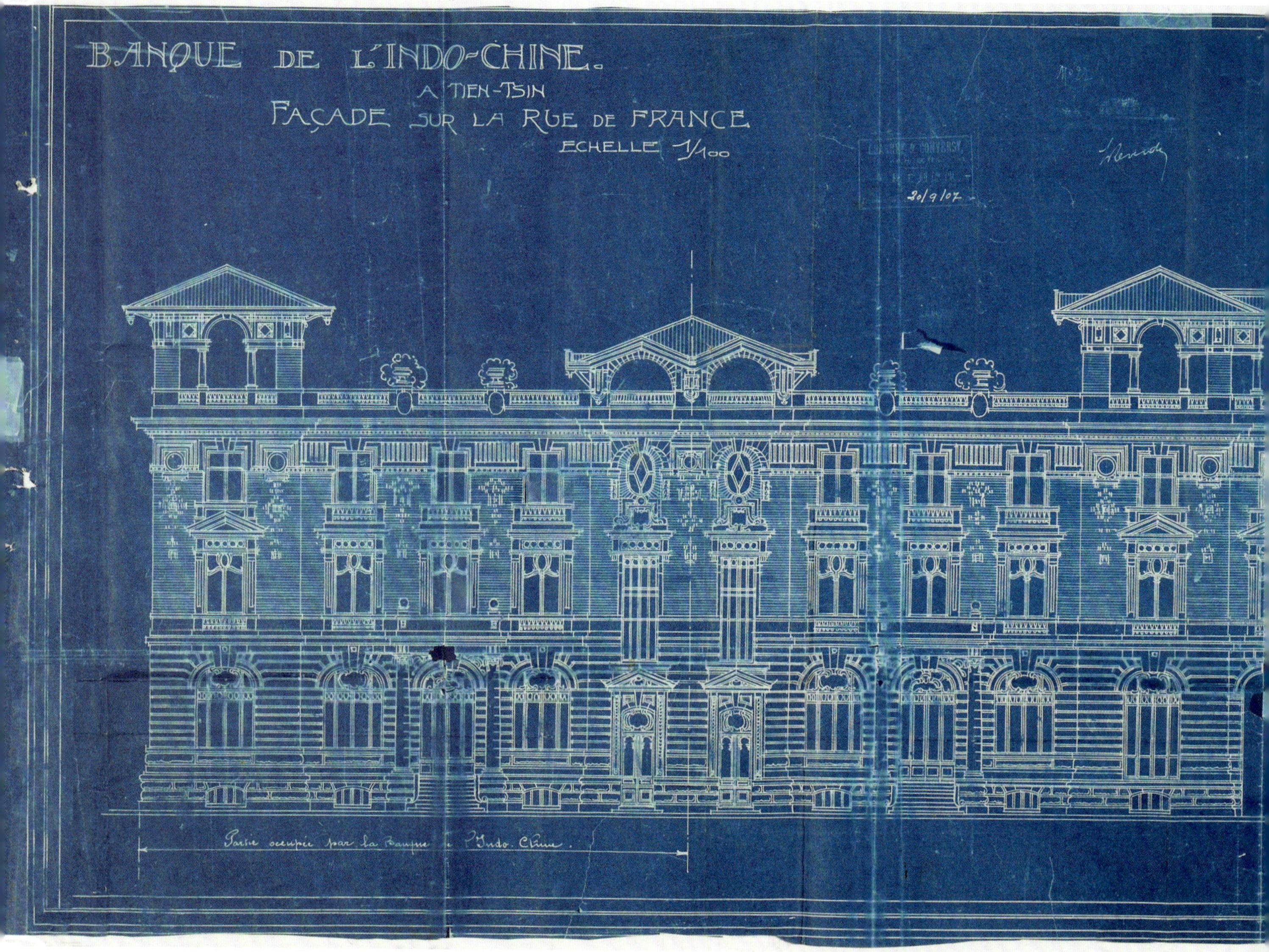

原设计图 Original blueprint

历史照片 Historic photo

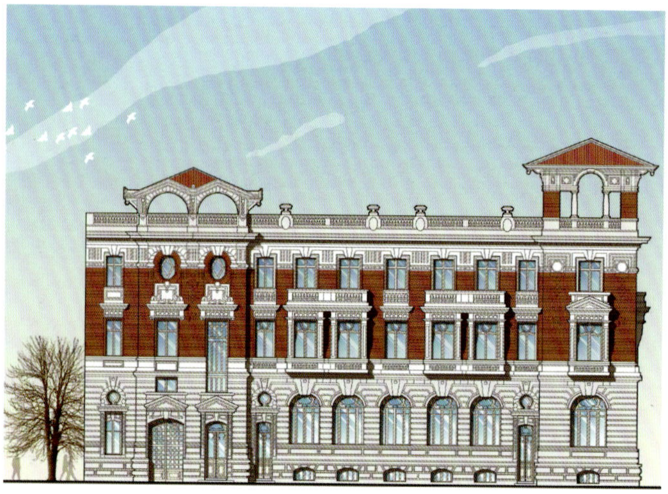

正立面渲染图 Facade Rendering

侧立面 Side elevation

The Former L'indo Bank of China was situated at 77~79 North Jiefang Road, Heping District. It is now the heritage site under city/county protection and the important protection historic architecture in Tianjin and used as an office building. The L'indo Bank of China was established in Paris in 1875. In the early twentieth century, branches of the bank were set up in Shanghai, Tianjin, Beijing, Hankou, Guangzhou, etc. The Tianjin branch was set in 1907 and specialized in inward documentary bills, outward documentary bills and foreign exchange trading. In 1913, the bank signed the Reorganization Loan Agreement together as the French member of the Consortia of Five Banks. After the Pacific War broke out, the L'indo Bank of China remained operating via political means and became the leading figure of the foreign banks in Tianjin. After the foundation of the People's Republic of China, the L'indo Bank of China was authorized to handle foreign exchange business. The bank was closed down in 1957. The office building of the L'indo Bank of China was designed by Credit Foncier d'Extreme Orient and built in 1912, covering an area of some 1,200 square meters and creating a floor area of about 3,500 square meters. The walls of the first floor were covers with stone-like concrete and those of the second and third floors were red brick and furred walls. It belongs to the Western classical architecture style.

侧入口 Side-entrance

整修前外景 Exterior view before renovation

主入口铁艺大门 Iron artwork of the main entrance door

外檐走廊 Corridor of the outer-eaves

内檐楼梯 Stairs

老暖气片 Old-fashioned heating devices

外檐局部 Partial outer-eaves

Former National Commercial Bank

原浙江兴业银行

浙江兴业银行成立于1907年，由浙江铁路公司创办，总行设在杭州，为“南四行”（浙江兴业、上海、新华、浙江实业）之一。天津分行1915年开业，经营存放款等业务，并发行钞票，1953年结业。该建筑位于和平区和平路237号，天津市文物保护单位，由华信工程司沈理源设计，建成于1922年，建筑面积约5 600 m^2。该建筑为带地下室的混合结构三层楼房，主立面为石材墙面。外檐装饰丰富，主入口柱廊中部为双柱，两侧为单柱，底层为塔什干柱式，二层为爱奥尼克柱式，庄严气派；立面三层窗户的装饰层层变异，严谨又丰富。室内装饰以精美的木雕和大理石雕刻。营业大厅呈圆形，用十四棵大理石圆柱环绕，上设环形梁，梁上雕有中国钱币图案，显示了建筑的功能特性和银行的实力。该建筑具有古典复兴建筑特征。

正立面渲染图 Facade Rendering

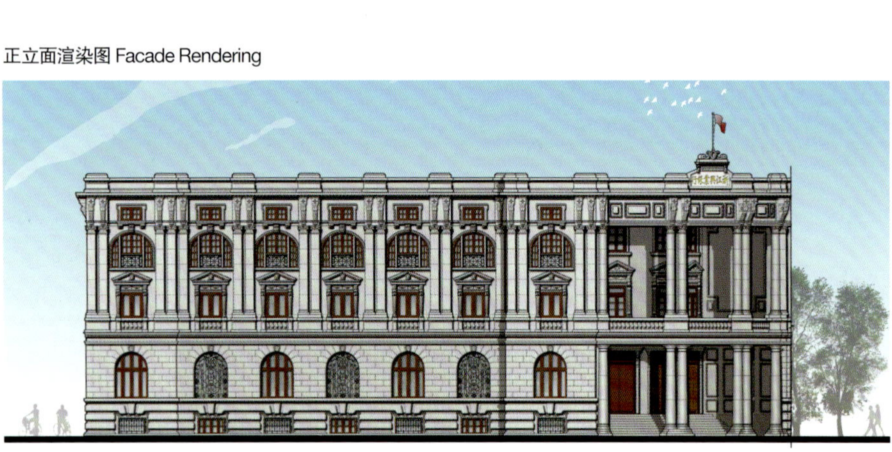

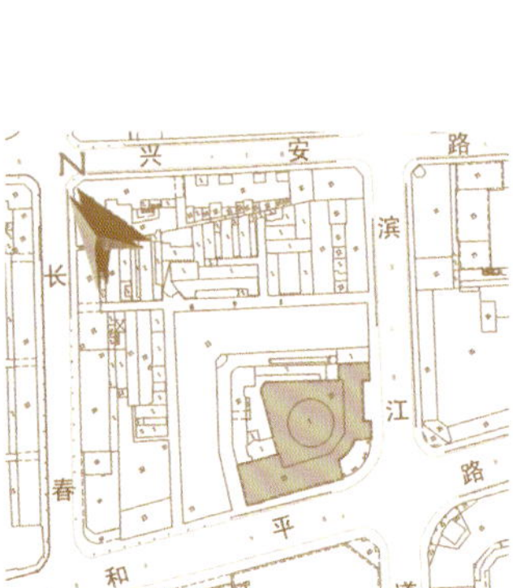

外檐檐口细部 Details of the Cornice

The Former National Commercial Bank was established by Zhejiang Railway Company in Hangzhou in 1907 and it was one of the "four southern banks" (Namely National Commercial Bank, Shanghai Bank, Sin Hua Trust Savings Bank, and Zhejiang Industrial Bank). Its Tianjin branch was opened in 1915 and specialized in deposits, loans and issuing bank notes. In 1953, the National Commercial Bank was closed down. Located at 237 Heping Road, Heping District, the building was designed by Shen Liyuan from Huaxin Engineering Company and was built in 1922. Listed as the heritage site under city protection. This three-storied mixed structure has a floor area of about 5, 600 square meters. The facade was constructed with stone material and the roof was beautifully decorated. Tashkent pillars were used on the first floor while Ionic pillars for the second floor. The ornaments used to decorate the windows vary from floor to floor. The indoor decorations are mainly exquisite carvings. The rotunda is surrounded by 14 marble columns and the roof is supported by annular girders engraved with China coins, which demonstrate the functions and power of the bank. It belongs to the classical Revival architectural style.

外檐大理石窗套 Marble window frame under the outer-eaves

外景 Exterior view

外檐局部 Partial outer-eaves

原设计图 Original blueprint

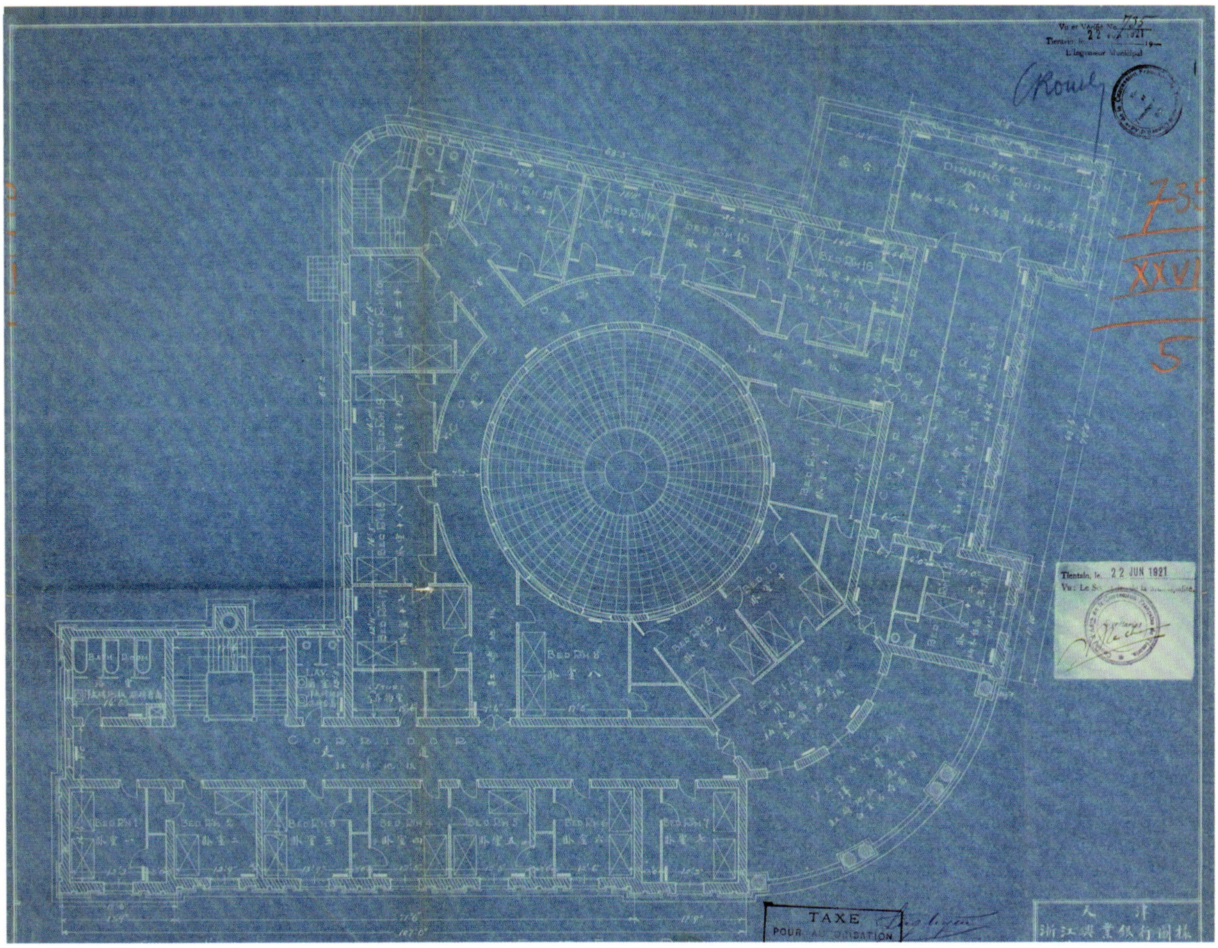

营业大厅 Business Hall

门套木雕 Wood-carvings door frame

二楼经理室精美木装饰
Elaborately wooden decoration in the manager’s office on the second floor

外檐穹顶局部 Details of the outer-eaves dome

营业大厅柜台大理石装饰 Marble-decorated counter in the business hall

Former Salt Industry Bank

原盐业银行

盐业银行创办于1915年，总行设在北京，是中国著名的商业银行，同年天津分行开业。初系盐务专业银行，后改招商股，成为普通商业银行，与金城、大陆、中南三家银行并称为"北四行"。该建筑位于和平区赤峰道12号，全国重点文物保护单位，特殊保护等级历史风貌建筑，建于1926年，现为中国工商银行，占地面积约2 100m²，建筑面积约6 200m²。华信工程司建筑师沈理源设计，四层砖混平顶楼房（设有地下室和中二层）。建筑主入口为通高的神庙式门厅，正立面用六根混合式巨柱支撑形成开敞长廊；墙面为水刷石饰面，饰以精美雕刻。建筑采用三段式构图，具有典型的古典主义特征，气势宏伟，装饰丰富，线脚细腻，富丽堂皇。

The same year when initially established, the Salt Industry Bank opened its branch in Tianjin in 1915. It was well-known as a commercial bank with its headquarter located in Beijing then. It was serving specially for salt affairs at the beginning and turned to an ordinary commercial bank later, being one of the four most famous in North China along with the Kinchen Bank, Continental Bank and China & South Sea Bank Ltd.

The site of the former Branch sits at 12 Chifeng Avenue, Heping District, which has been listed as one of the important heritage site under state protection and the special protection historic architecture in Tianjin. Now, the building is used by Industrial and Commercial Bank of China, covering an area of about 2,100 square meters, and about 6,200 square meters in floor area.

As one of the splendid works of Mr. Shen Liyuan, a Chinese architect, the building was originally built in 1926. The four-storied brick-concrete architecture is composed classically of three sections with its basement and intermediate-layer on the second floor as well. It looks more magnificent for its extractive designs: Main Entrance formed in highly fane-style, facade colonnade structured by six massive columns, walls faced with granite-plaster, and the whole building decorated with delicate architrave, rich carvings and elegant architraves.

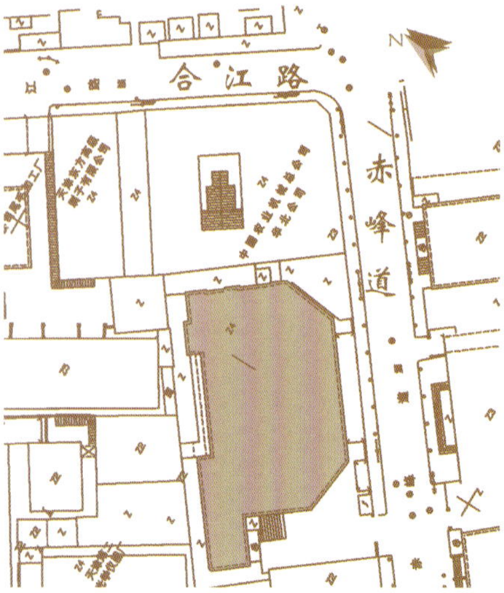

历史照片 Historic photo

主入口外景 Main Entrance

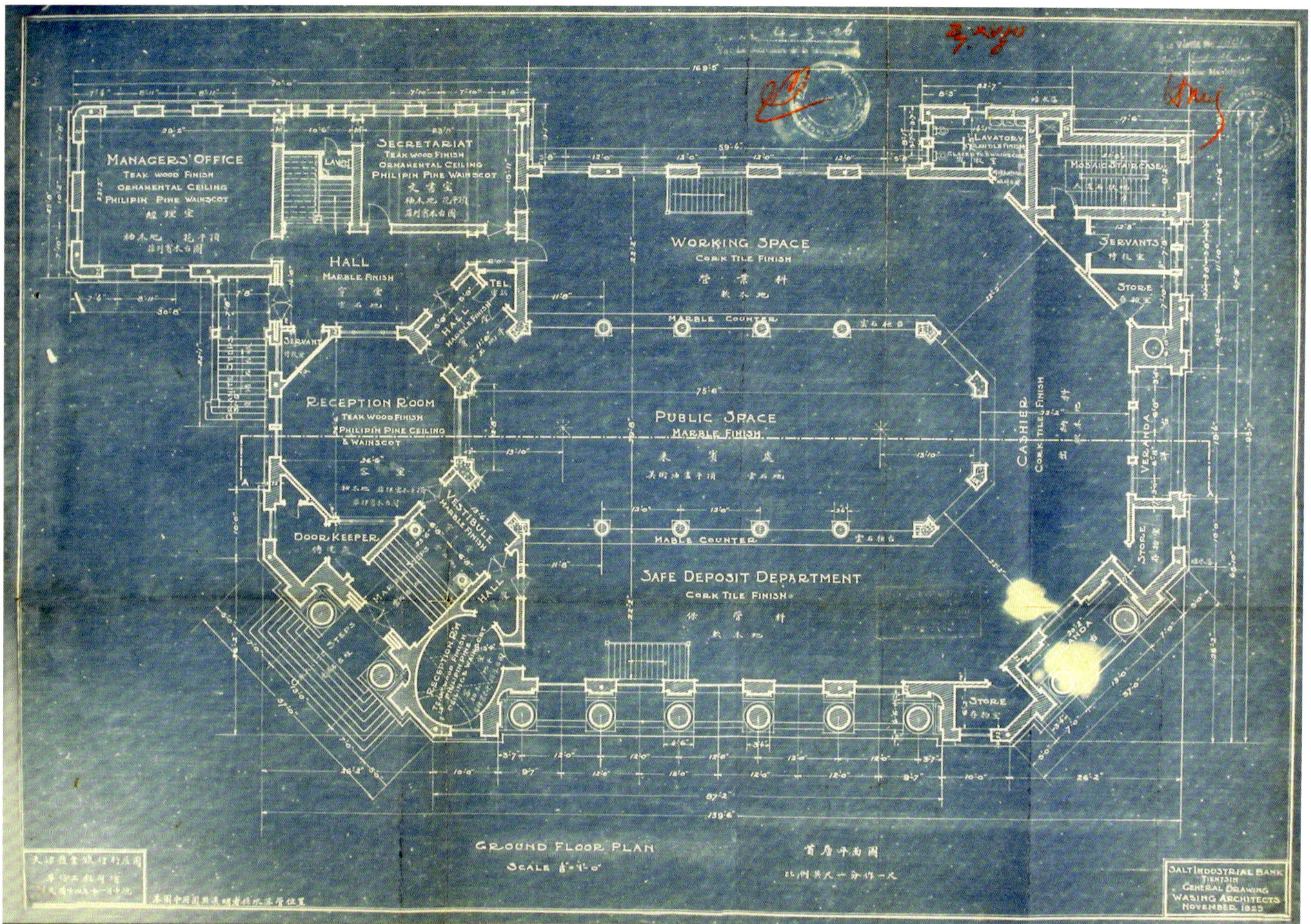

原设计图 Original blueprint

外景局部 Partial outer-eaves

营业大厅 Business Hall

世界时钟 WORLD TIME CLOCK

三层大厅 Lobby on third floor

老家具Old-fashioned furniture

老暖气罩Old-fashioned radiator enclosure

会议室 Meeting Room

室内西门子老钟表Old-fashioned Siemens clock

室内老吊灯 Old-fashioned lamp

三层门厅 Lobby on third floor

楼梯雕花细部 Decorations on the stairs

内檐彩色玻璃花窗 Stained-glass window under the inner-eaves

外景 Exterior view

Former China & South Sea Bank Ltd.

原中南银行

原中南银行位于和平区解放北路88号，天津市文物保护单位，特殊保护等级历史风貌建筑，现为中国建设银行。中南银行成立于1921年，为印尼华侨黄奕柱和原交通银行北京分行经理胡笔江共同创办的私营股份制商业银行，总行设于上海，天津分行于1922年开业，主营存放款业务。银行定名为中南银行，是取中国与南洋华侨合作之意。因该行为华侨投资，经北洋政府特批，允许其发行钞票。二十世纪三、四十年代，该行又陆续开办了外汇及信托证券业务。1952年中南银行结业。该建筑建于二十世纪初，占地面积约3 500 m²，建筑面积约3 200 m²。原为二层钢混结构楼房（带地下室），1938年由华信工程司沈理源设计增建一层。建筑正立面为石材饰面，规整统一，立面设计遵循古典主义的对称原则，设置贯通两层的半柱，柱身有爱奥尼克柱式花纹，但省略了柱头。入口门头细致婉约，与粗犷的主体建筑形成有趣的对比。室内装修讲究，功能布局合理，木地板、木楼梯等设施保存完好。

穹顶Dome

外檐侧立面入口台阶
Stairs at the side elevation' s entrance

主入口Main Entrance

The China & South Sea Bank Limited was established as a commercial bank in 1921 in Shanghai. It was founded by Huang Yizhu who was overseas Chinese from Indonesia and Hu Bijiang, manager of the Beijing Branch of the Bank of Communications. In 1922, its Tianjin branch was opened in 88 North Jiefang Road in Heping District to provide deposit and loan services. Now the building is the heritage site under city protection and the special protection historic architecture in Tianjin. The bank was named the China & South Sea Bank because it was established under the joint efforts of China and overseas Chinese from Indonesia. It was authorized by the Beiyang Government to issue banknotes for its overseas Chinese investment. In the 1930's and 1940's, foreign exchange business and trust and securities were started in the Bank. The bank was closed down in 1952.

The two-storied reinforced concrete building with a basement was built in the early 1900's and its floor space was 3,500 square meters while its floor area 3,200 square meters. In 1938, the third floor was added to the original building by Huaxin Engineering Company following Shen Liyuan's design. The facade of the building was decorated with stone in perfect symmetry and pillars bearing Ionic patterns were used to support the building. The delicately-designed door and the main body of the building are of totally different styles. The rooms were well-designed and beautifully decorated. Up to now, the wood floor and the flights are in good condition.

外檐背立面 Exterior view of the back side elevation

外景 Exterior view

积金管理中心
银行
Bank

内檐门 Gate under the inner eaves

内檐楼梯 Stairs under the inner eaves

共享大厅 Sharing hall

历史照片 Historic photo

Former Kinchen Bank

原金城银行

原金城银行大楼位于和平区解放北路108号，天津市文物保护单位，重点保护等级历史风貌建筑，现为办公用房。金城银行开业于1917年，由中国近代银行家周作民创办，发起人为倪嗣冲、任凤苞、胡笔江等，行名取“金城汤池永久坚固之意”，总行设于天津，1936年迁往上海。金城银行是近代中国著名的民族资本商业银行，也是北四行（金城、盐业、大陆、中南）的核心和支柱。其主要业务为存款、放款及投资，自营企业有经营商贸的通成公司、专营公债买卖的丰大号，以及后来成为华北地区华商保险业巨擘的太平洋保险公司。1952年加入金融业全行业公私合营后宣告结束。

金城银行大楼建于1937年，占地面积约4 500m^2，建筑面积约3 200m^2，由中国第一代著名建筑师——华信工程司沈理源先生设计。建筑为二层砖木结构，带地下室，外墙装饰面砖。建筑布局规整对称，造型为古典复兴与民居建筑的混合，立面装饰丰富，大楼原有高耸多坡可利用屋顶，在1976年地震中震损。在2008年迎奥运环境综合整治工作中，由建筑使用单位对大楼进行了整修，恢复了历史原貌。

整修前外景 Exterior view before renovation

外景 Exterior view

Situated at 108 North Jiefang Road, Heping District, the former Kinchen Bank is now the heritage site under city protection and the important protection historic architecture and is currently used as an office building. Kinchen Bank was established in Tianjin in 1917 by Zhou Zuomin, Ni Sichong, Ren Fengbao and Hu Bijiang. In 1936, the bank was relocated to Shanghai. Kinchen Bank was the core of the "four northern banks" in China (the other three were Salt Industry Bank, Continental Bank and China & South Sea Bank) in the 1920s. The major services of Kinchen Bank were deposit, loan and investment and it also ran several companies including Tongcheng Company, Fengdahao and the Pacific Insurance Agency which became the leading figure in China's insurance industry. In 1952, it was grouped into the "Joint Office of Joint Public-Private Banks" with other eight Chinese banks. The Kinchen Bank building was built by Huaxin Engineering Company in 1937 following the design of Shen Liyuan, China's first-generation designer and it covers an area of 4,500 square meters and its floor area was 3,200 square meters. This two-storied brick-wood building with a basement was laid out in perfect symmetry and its exterior walls were covers with decorative tiles. As it combined the classical revival style and vernacular dwelling, its facade is beautifully decorated and it has towering roofs. In the 1976 Tangshan Earthquake, the roofs were damaged and later the building was restored before the advent of the Beijing 2008 Olympic Games.

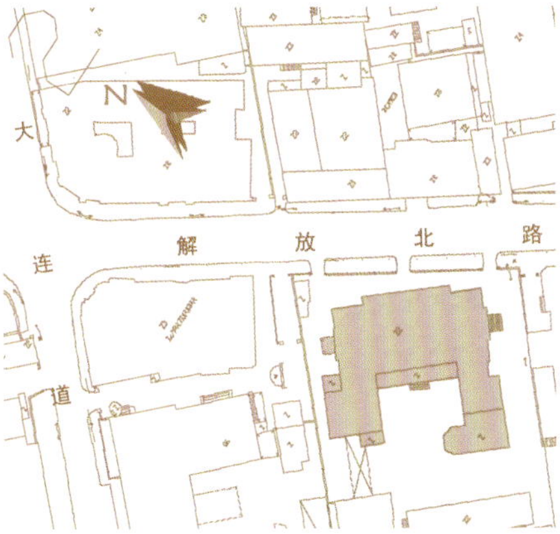

外景 Exterior view

外檐局部 Partial outer-eaves

主入口 Main Entrance

围墙 Enclosure

彩色玻璃与楼梯 Stained glass window and stairs

内檐楼梯 Stairs under the inner-eaves

铜制钥匙孔 Copper keyhole

护墙板 Wall-panels

↓壁炉 Fireplace

Former Continental Bank

原大陆银行

原大陆银行位于和平区哈尔滨道70号，天津市区、县文物保护单位，重点保护等级历史风貌建筑，现为交通银行。大陆银行成立于1919年，总行设在天津，是近代中国著名的私营股份制商业银行——“北四行”之一，股东有民国总统冯国璋、江苏督军李纯、江苏省省长齐耀琳等。主要经营保管、信托、买卖股票及有价证券业务。该行曾建设仓库经营仓储业，打破了英商平和洋行对天津仓储业的垄断。1953年大陆银行结业。该建筑1921年建成，占地面积约2 000 m^2，建筑面积约4 900 m^2，由基泰工程司设计。建筑为砖混结构三层楼房，带地下室，外檐为混水墙面，饰以简易的爱奥尼克半壁柱，屋顶设有塔楼。室内布局合理，装饰考究，为折中主义建筑风格。

正立面渲染图 Facade Rendering

The Continental Bank was opened in 1919 in Tianjin and located at 70 Haerbin Avenue, Heping District. Now it is the office building of the Bank of Communications and it is the heritage site under city/county protection and the important protection historic architecture in Tianjin. It was one of the "four northern banks" and invested by Feng Guozhang, the President of the Republic of China, Li Chun, the Jiangsu Military Governor and Qi Yaolin, the then Governor of Jiangsu Province. The core business of the bank was in trust, stocks and securities and it had invested in warehousing and broke up British Pinghe Company's monopoly of the warehousing industry in Tianjin. The Continental Bank was closed down in 1953. The office building of the Continental Bank was built in 1921. It covers an area of some 2,000 square meters and its floor area was about 4,900 square meters. This three-storied brick-concrete building with a basement was designed by Jitai Engineering Company and belongs to eclectic architecture. The exterior furred wall was decorated with Ionic pilaster block and a tower was built on the roof. The rooms are well laid out and beautifully decorated, bearing the characteristics of eclectic architecture.

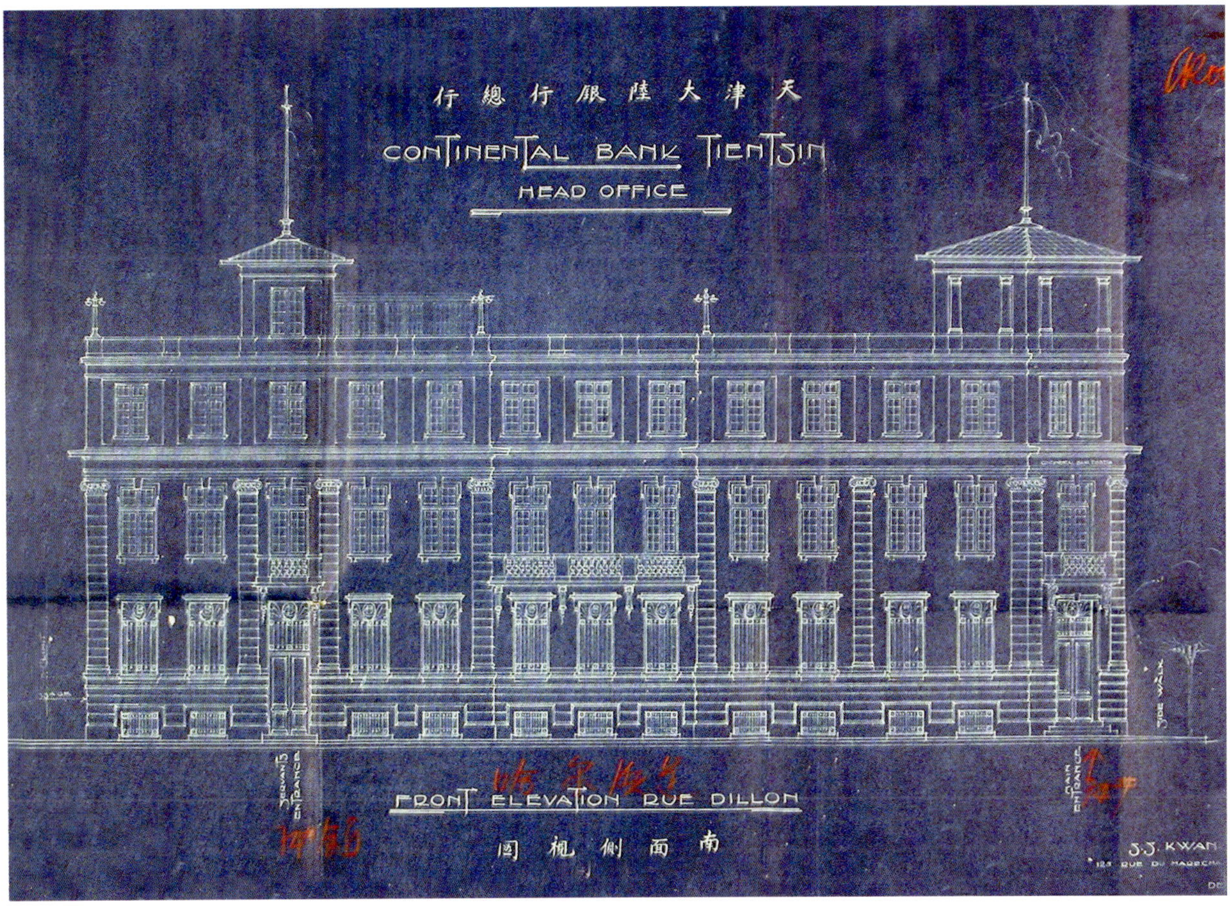

原设计图 Original blueprint

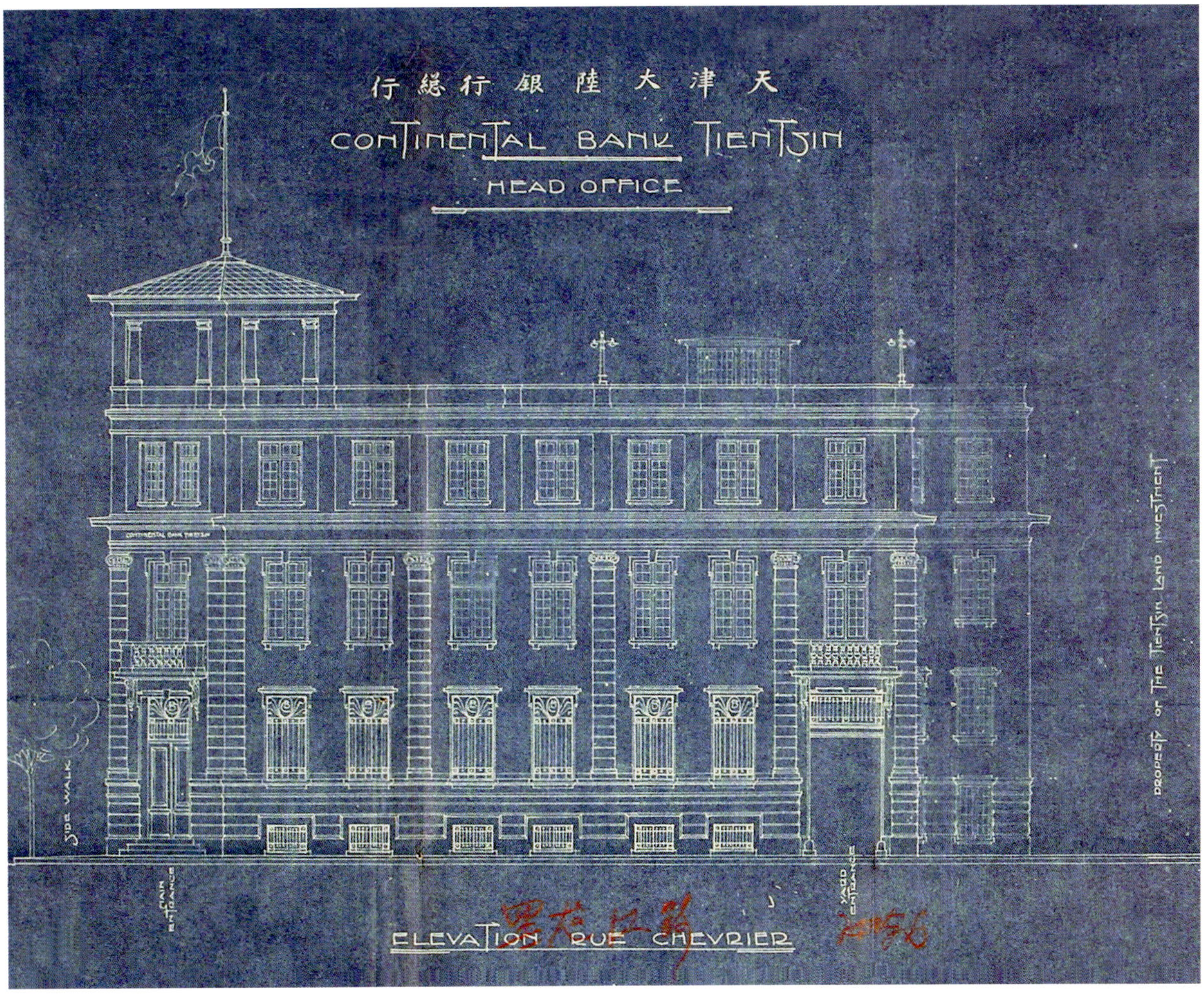

外景 Exterior view

老吊灯 Old-fashioned lamp

二楼过厅内景
Interior scenery in the corridor of the second floor

内檐护墙板局部雕花
Decoration on the wall-panels

内檐楼梯 Stairs under the inner-eaves

夜景 Night view

Former Location of Joint Savings Society Bank

原四行储蓄会天津分会

四行储蓄会由盐业、金城、中南、大陆四家中资银行联合设立，1923年开业，总会设于上海，在上海、天津、汉口分别设立分会。四行储蓄会为会员制储蓄机构，四行为基本会员，储户为一般会员。该会提倡社会各界正当储蓄，积极吸揽储金，用于投资，并由四行共同担负保本付息之责。原四行储蓄会天津分会位于和平区解放北路147号，天津市区、县文物保护单位，特殊保护等级历史风貌建筑，现为中国工商银行。该建筑建于1923年，占地面积约1 400 m^2，建筑面积约3 000 m^2。混合结构三层楼房（带地下室），外檐为清水砖墙。建筑构图庄严稳重，布局简洁规整，立面层次丰富、装饰精美，正入口的高台阶及上方四棵贯通二、三层的爱奥尼克列柱，使整个建筑气势恢宏，具有简洁的仿希腊古典复兴建筑特征。

外檐局部 Partial outer-eaves

The Joint Savings Society Bank was established in the joint endeavor of the "four northern banks", namely, Kinchen Bank, Salt Industry Bank, Continental Bank and China & South Sea Bank. Opened in 1923 and headquartered in Shanghai, the bank set up branches in Shanghai, Tianjin and Hankou. The "four northern banks" were its basic members and the depositors were its general members. The society advocated savings from all walks of life and made great effort to collect savings to invest and the four banks ensure to repay the capital and interest to depositors. The Joint Savings Society Bank was situated at 147 North Jiefang Road, Heping District, which is now the office building of the Industrial and Commercial Bank of China, and has been listed as the heritage site under city/county protection and the special protection historic architecture in Tianjin. This three-storied mixed structure building with a basement was built in 1923 and it covers an area of some 1,400 square meters and has a floor area of about 3,000 square meters. Dry brick exterior walls made the building majestic and the four Ionic colonnades made it magnificent. It is a copy of the revival of classical Greece style.

营业大厅木雕柜台 Wood-carved counter in the Business Hall

外景 Exterior view

Former Tung Lai Bank

原东莱银行

原东莱银行位于和平区和平路287号，天津市文物保护单位，重点保护等级历史风貌建筑，现为办公用房。东莱银行为中资银行，成立于1918年，1919年设立天津分行。该行经营存、贷款及兑换外币、有价证券买卖等业务，1953年结业。该建筑建于1930年，占地面积约3 300 m^2，建筑面积约7 500 m^2，由贝伦特工程司设计，大楼建成后东莱银行从宫北大街旧址迁入新楼。该建筑为混合结构三层楼房，局部四层，混水墙面。主入口两侧为贯通两层的科林斯柱廊，檐部为山花，具有典型的仿希腊古典复兴建筑特征。三层为阁楼层，配方柱廊，入口处的阁楼层上设高耸的重檐圆形塔楼，使建筑主入口突出，同时建筑形式也增加了折中主义的成分。室内装饰考究，营业大厅内为彩色水磨石地面、大理石护墙板，充分显示了银行的实力。

历史照片 Historic photo

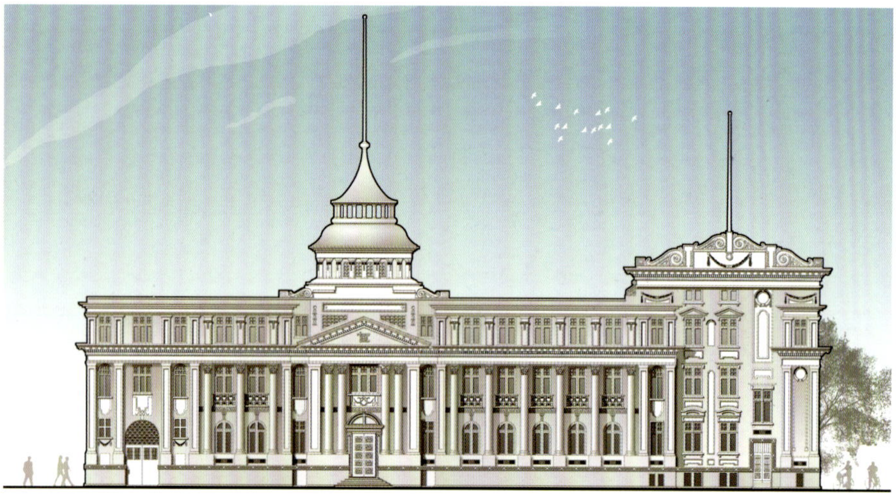

正立面渲染图 Facade Rendering

Former Tung Lai Bank, a Chinese-funded bank established in 1918, had its Tianjin branch set up at 287 Heping Road, Heping District in 1919. Its former site is now the heritage site under city protection and the important protection historic architecture in Tianjin and is used as an office building. The bank mainly engaged in deposit, loan, exchange of foreign currency and securities dealing. It was closed down in 1953. The office building of Tung Lai Bank was designed and built by Behrendt Engineering Company in 1930. It covers an area of some 3,300 square meters and has a floor area of about 7,500 square meters. This three-storied mixed structure with part of the building being four-storied has plastered brick walls, Corinthian colonnades at the main entrance and pediment at the top of the front which belongs to the revival of classical Greece style. Meanwhile, elements of eclecticism were also added to it as a multiple-eave round tower was built on the attic over the main entrance to give prominence to the main entrance. Besides, the colored terrazzo floor and marble wall panel demonstrated the financial power of the bank.

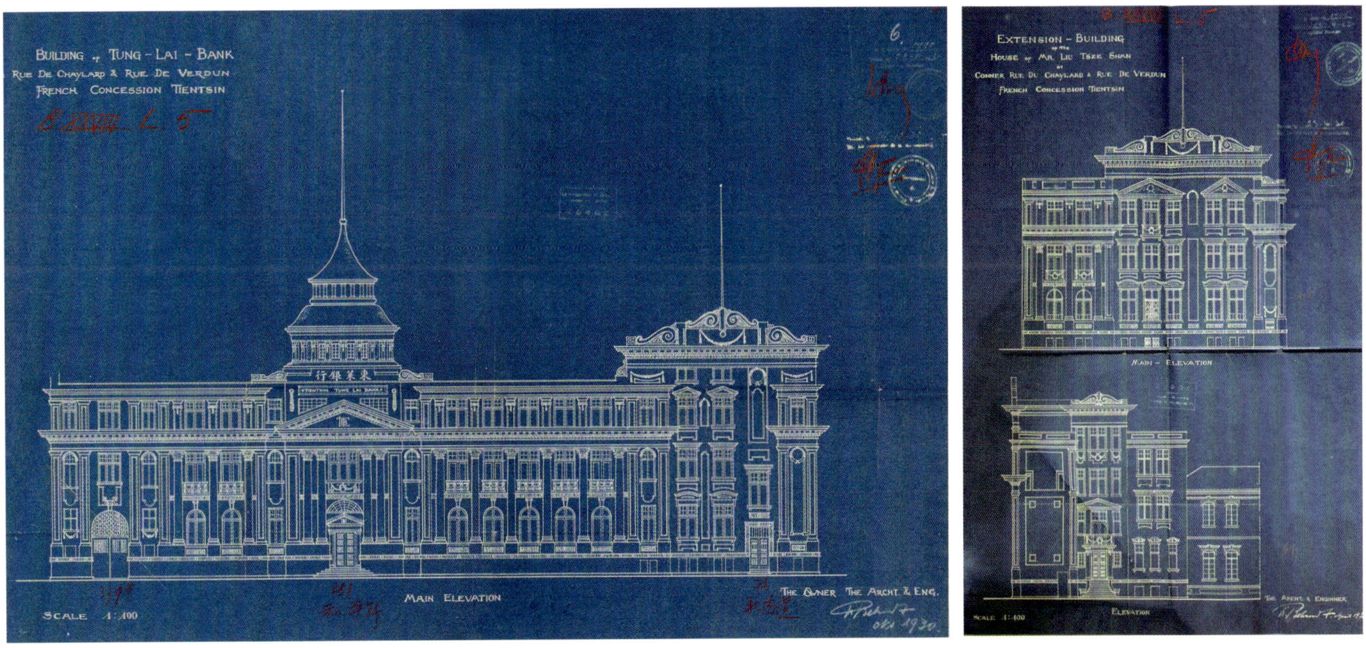

原设计图 Original blueprint

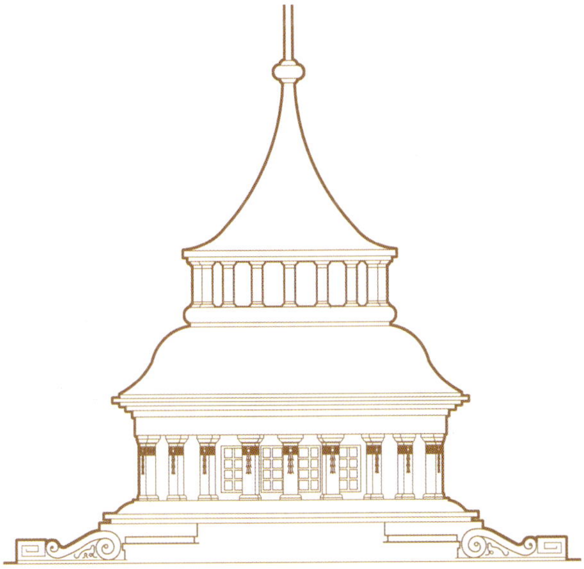

外景 Exterior view

科学宫
TIAN JIN CITY COMMERCIAL BANK
S & T SUB BRANCH

顶棚 Ceiling

顶棚 Ceiling

内檐木制天花板 Wooden ceiling

贵宾厅 Hall for the honored guests

内檐旋转楼梯 Revolving stairs

木雕细部 Details of the carved wood structure

楼梯间内景 Inner view of staircase

Former Sin Hua Trust Savings Bank

原新华信托储蓄银行

原新华信托储蓄银行位于和平区解放北路10号，天津市文物保护单位，重点保护等级历史风貌建筑，现为办公用房。新华信托储蓄银行由中资中国银行、交通银行于1919年联合创办，原名新华储蓄银行，1931年改组后更名为新华信托储蓄银行。该行以“提倡社会储蓄，培养人民俭德”为宗旨，天津分行主要业务为存放款及储蓄，兼营经收信托款项、买卖有价证券、房地产、保险等业务，并发行流通储蓄券，后又推出整借零还小额贷款，首创人寿储金、生活储金、教育储金。据原中国人民银行参事室主任、中国银行界元老沈日新老人回忆，新华信托储蓄银行当时还有一项重要的业务，即所有的中资银行每天下午四点均要到该行来进行汇总结账，沟通情况，可以说是天津中资银行的大本营。1952年该银行结业。该建筑建于1934年，占地面积约1 100 m^2，建筑面积约9 200 m^2，由华信工程司建筑师沈理源设计，正门设于解放北路与滨江道转角处，营业大厅贯通一、二两层。主体六层，局部八层，钢筋混凝土框架结构，外檐石材饰面，建筑造型庄重挺拔，强调竖向构图，整体感很强，是二十世纪诞生的现代建筑的典型代表。

正立面渲染图 Facade Rendering

外景 Exterior view

外檐细部 Details of the outer-eaves

屋顶塔楼 Tower on the roof

↓外檐细部 Details of the outer-eaves

The Former Sin Hua Trust Saving Bank, situated at 10 North Jiefang Road, Heping District, is now the heritage site under city protection and the important protection historic architecture in Tianjin and used as an office building. The Sin Hua Trust Savings Bank, previously Sin Hua Savings Bank, was established by Chinese-funded Bank of China and Bank of Communications in 1919 and renamed as the Sin Hua Trust Savings Bank after its reorganization in 1931. The bank advocated a savers' society. Its Tianjin branch first engaged in mainly deposit, loan, savings, trust payment, securities dealing, real estate and insurance. Later it issued savings certificates, promoted a service of petty loan, and initiated life insurance savings, life savings and education savings.

Shen Rixin, one of the earliest financial experts in China, head of the advisory office of the People's Bank of China recalled that the Sin Hua Trust Savings Bank was the base camp of all the Chinese-funded banks in Tianjin because officers from all the banks got together to

二楼回廊 Cloister on second floor

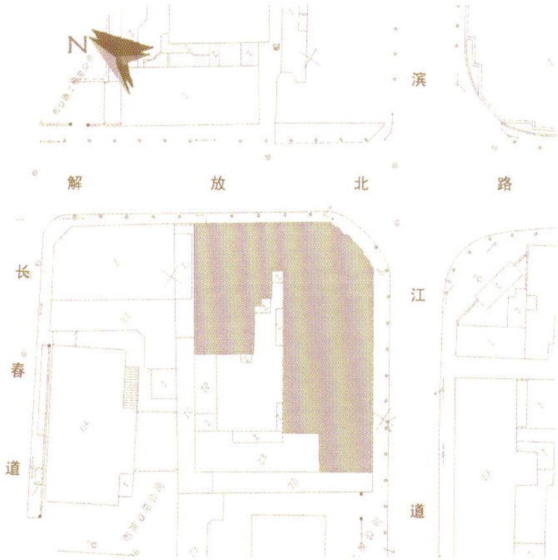

settle account and exchange information everyday at 4 o'clock. In 1952, the Sin Hua Trust Savings Bank was closed down.

The building was designed by Shen Liyuan and built by Huaxin Engineering Company in 1934. It covers some 1,100 square meters and has a floor area of about 9,200 square meters. The main entrance was at the corner of North Jiefang Road and Binjiang Avenue. The business hall was two-storied high and the main body of the reinforced concrete building is six-storied while part of it is eight-storied. The stone-decorated facade had highlighted the vertical format and made the building a classic representative of modern architectures in the twentieth Century.

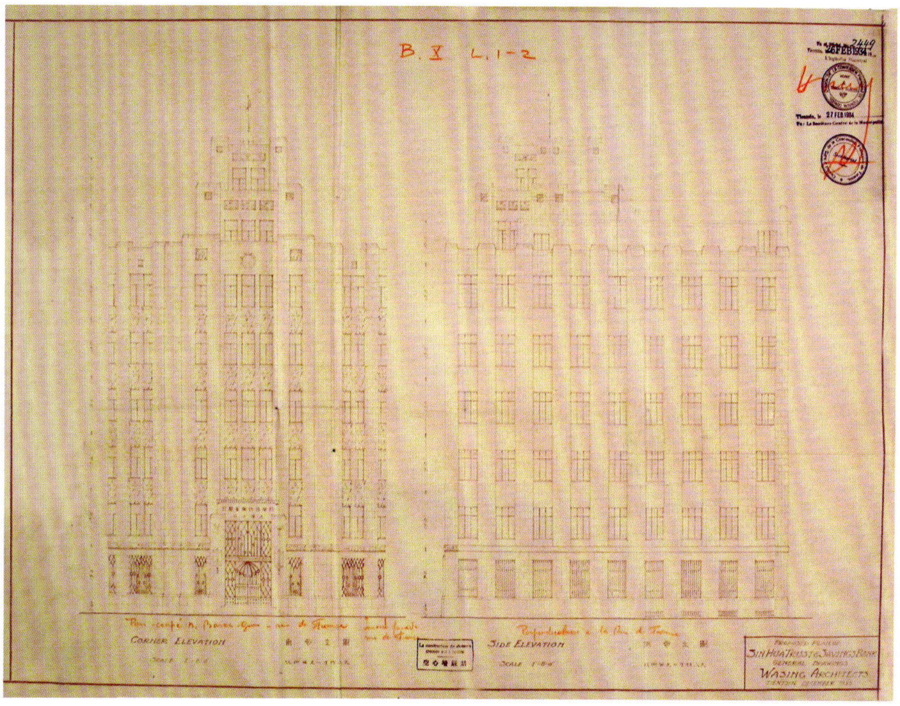

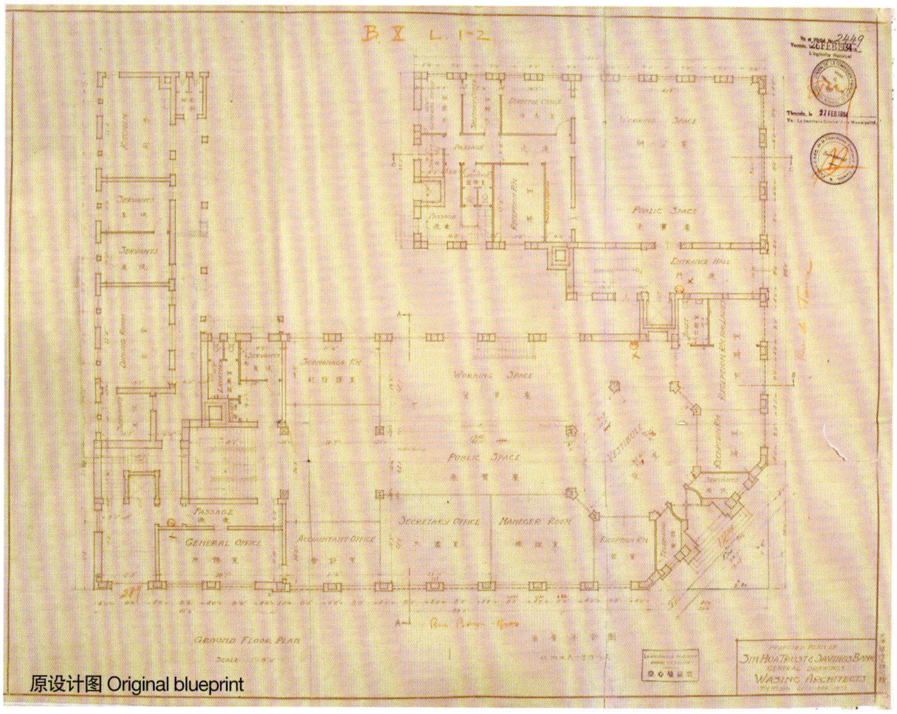

原设计图 Original blueprint

Former Central Bank

原中央银行

原中央银行位于和平区解放北路117~119号，天津市文物保护单位，特殊保护等级历史风貌建筑，现为中国人民银行。中央银行为1928年在南京成立的国家银行，主要业务为调剂金融，经营贷款、贴现业务，代理国库收存关税、盐税，并发行钞票。天津分行于1931年开业。该建筑于1926年建设，占地面积约2 400m²，建筑面积约3 100m²，原为中日合资的中华汇业银行，中央银行天津分行于1936年购得此楼。该建筑由华信工程司建筑师沈理源设计。为三层混合结构楼房，设有半地下室。外檐正立面用四棵爱奥尼克巨柱和两棵方柱构成对称立面，造型庄重大方，布局严谨，室内装修华丽考究，具有古典复兴主义建筑特征。

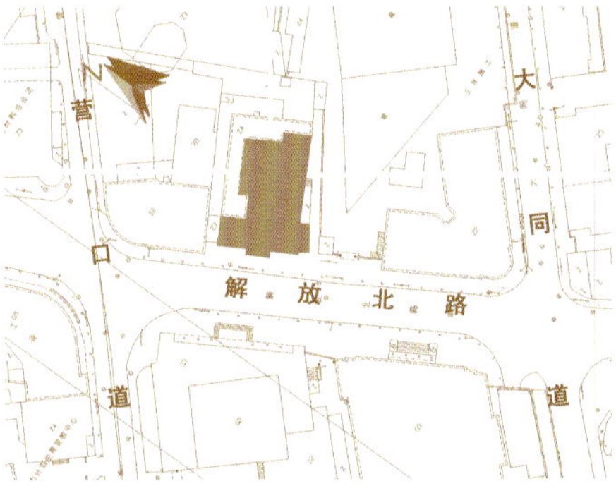

The former Central Bank was situated at 117~119 North Jiefang Road, Heping District. It is now the heritage site under city protection and the important protection historic architecture in Tianjin and used as the office building of the Bank of China. The Central Bank of China was established in Nanjing in 1928, specializing in supervising financial affairs, loan operation, discount, collecting tariff, salt tax and issuing bank notes. The Tianjin Branch of the Central Bank was opened in 1931. This building was constructed in 1926, covering about 2,400 square meters and producing about 3,100 square meters floor area. The building, used to be owned by the Sino-Japan co-invested China Banco Delta Ásia, was bought by the Tianjin Branch of the Central Bank in 1936. This three-storied mixed structure with semi-basement was designed by Shen Liyuan from Huaxin Engineering Company. With four Ionic pillars and two square columns standing in symmetry in the facade, the building appears majestic and magnificent. The rooms were well laid out and beautifully decorated, epitomizing the classical architectural style.

历史照片 Historic photo

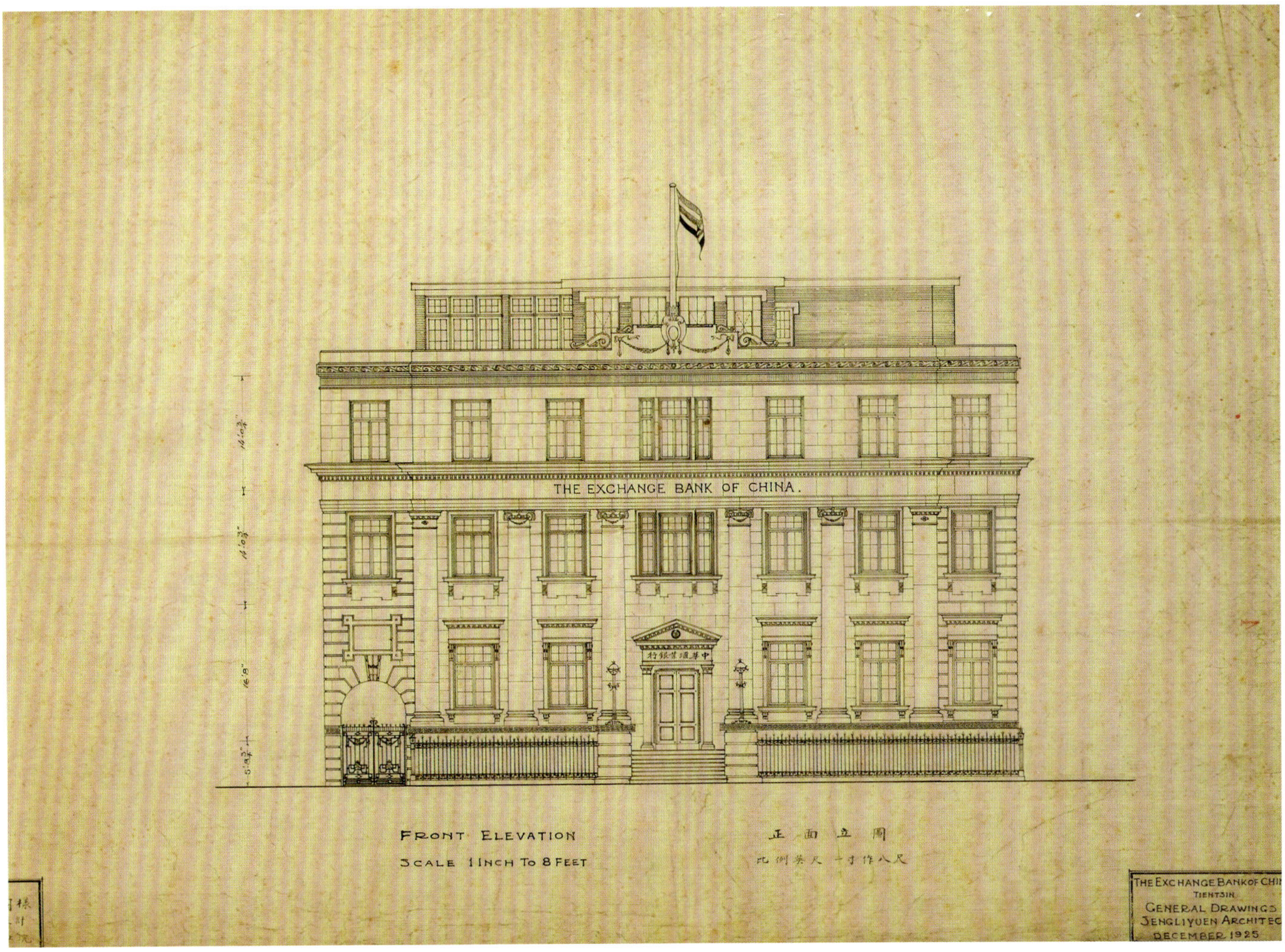

原设计图 Original blueprint

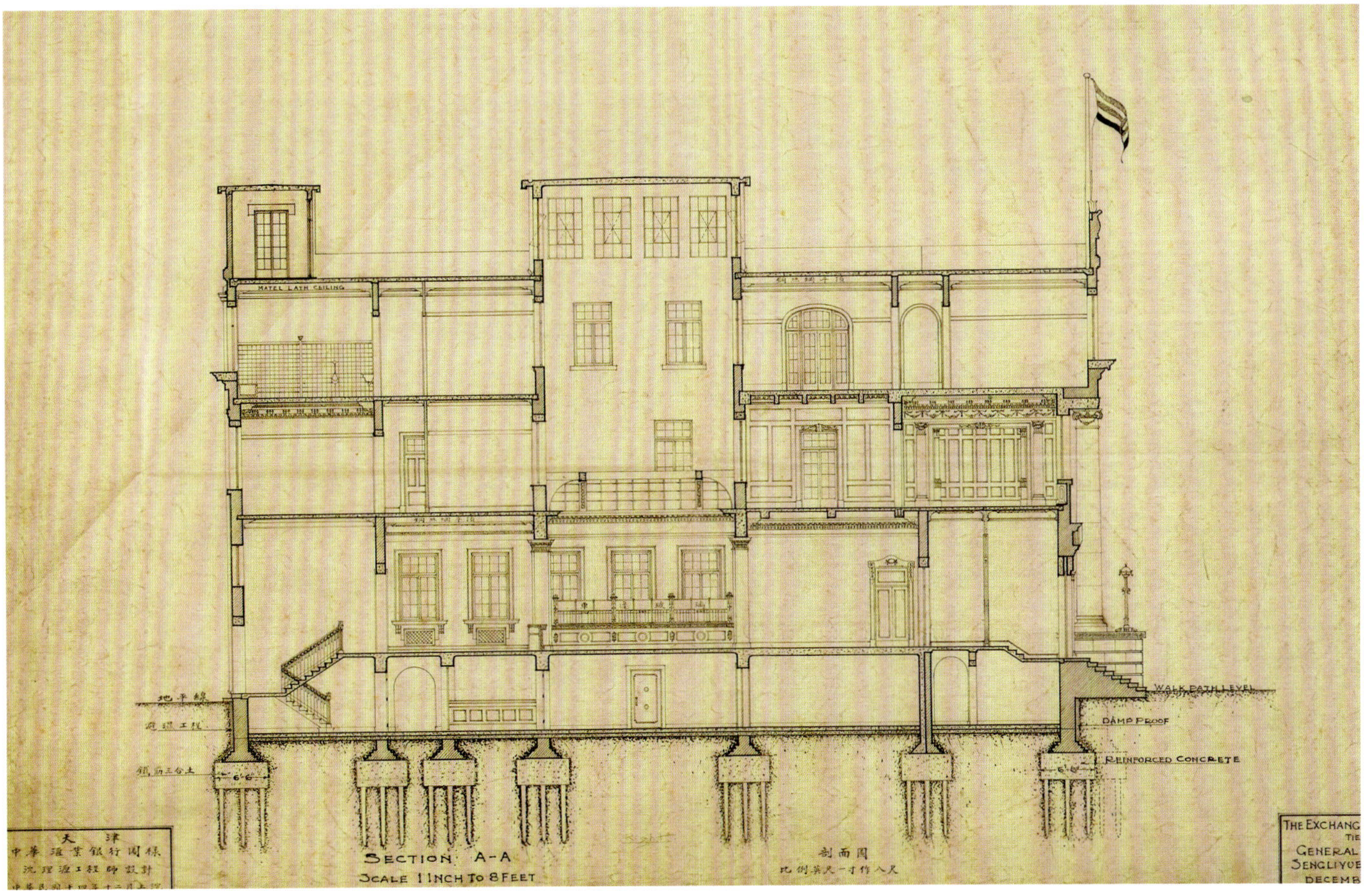

外景 Exterior view

二楼大厅 Lobby on second floor

↓会议室 Meeting Room

内景 Inside view

Former Industrial Bank of China

原中国实业银行

原中国实业银行位于和平区大同道15号，重点保护等级历史风貌建筑，现为办公用房。该行为实业家周学熙发起创办，1919年4月开业，总行设在天津，筹建时除募集股外，由北洋政府指定长芦、东纲及两淮盐商投资，旨在为周学熙所掌握的滦州矿务局、启新洋灰公司、耀华玻璃公司、华新纺织公司解决大量流动资金的存储及转帐。1934年被宋氏家族吞并。该建筑建于1921年，基泰工程司设计并监造。占地面积约3 000 m^2，建筑面积约3 000 m^2。为带地下室的混合结构二层楼房，平屋顶。正立面一至二层中部为混水墙面，石材饰面，其他立面为红砖清水墙面，转角处设水刷石饰面的抱角石。正立面以六棵爱奥尼克柱式承托檐部，形成开敞柱廊，柱廊上部为高耸的女儿墙，柱及檐部、窗套均为水刷石饰面。首层窗套用山花造型装饰。建筑简约大气，具有古典复兴建筑特征。

外景 Exterior view

Situated at 15 Datong Avenue, Heping District, the former Industrial Bank of China is now an important protection historic architecture in Tianjin and is used as an office building. In April, 1919, the Industrial Bank of China was established by Zhou Xuexi, a modern industrialist in Tianjin. Apart from the funds raised, the Beiyang Government appointed Changlu Salt Company and Donggang Salt Company to invest in the bank to provide circulating fund and to transfer accounts for the enterprises of Zhou Xuexi Industry Group including Luanzhou Mining Bureau, Qixin Cement Company, Yaohua Glass Company and Huaxin Textile Company. In 1934, the Industrial Bank of China was taken over by the Soong Family. The office building of the Industrial Bank of China was designed and built by Jitai Engineering Company in 1921. This two-storied flat-roofed building with a basement covers about 3,000 square meters and has a floor area of some 3,000 square meters. Most of its facade was furred wall decorated with stone and other exterior walls were red brick walls. At the corner, there is a corner stone made of granitic plaster. The six Ionic pillars supporting the roof formed an open colonnade topped with high-rise parapet. The pillars, roof and window trims are decorated with granitic plaster and pediments are used to the window trims of the first floor. It belongs to the classical Revival architecture style.

历史照片 Historic photo

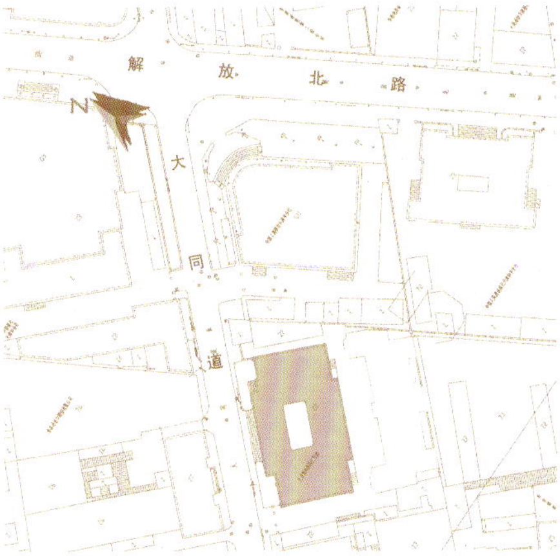

外檐门
Gate under the outer-eaves

外檐窗
Windows below the outer-eaves

正立面渲染图 Facade Rendering

原设计图 Original blueprint

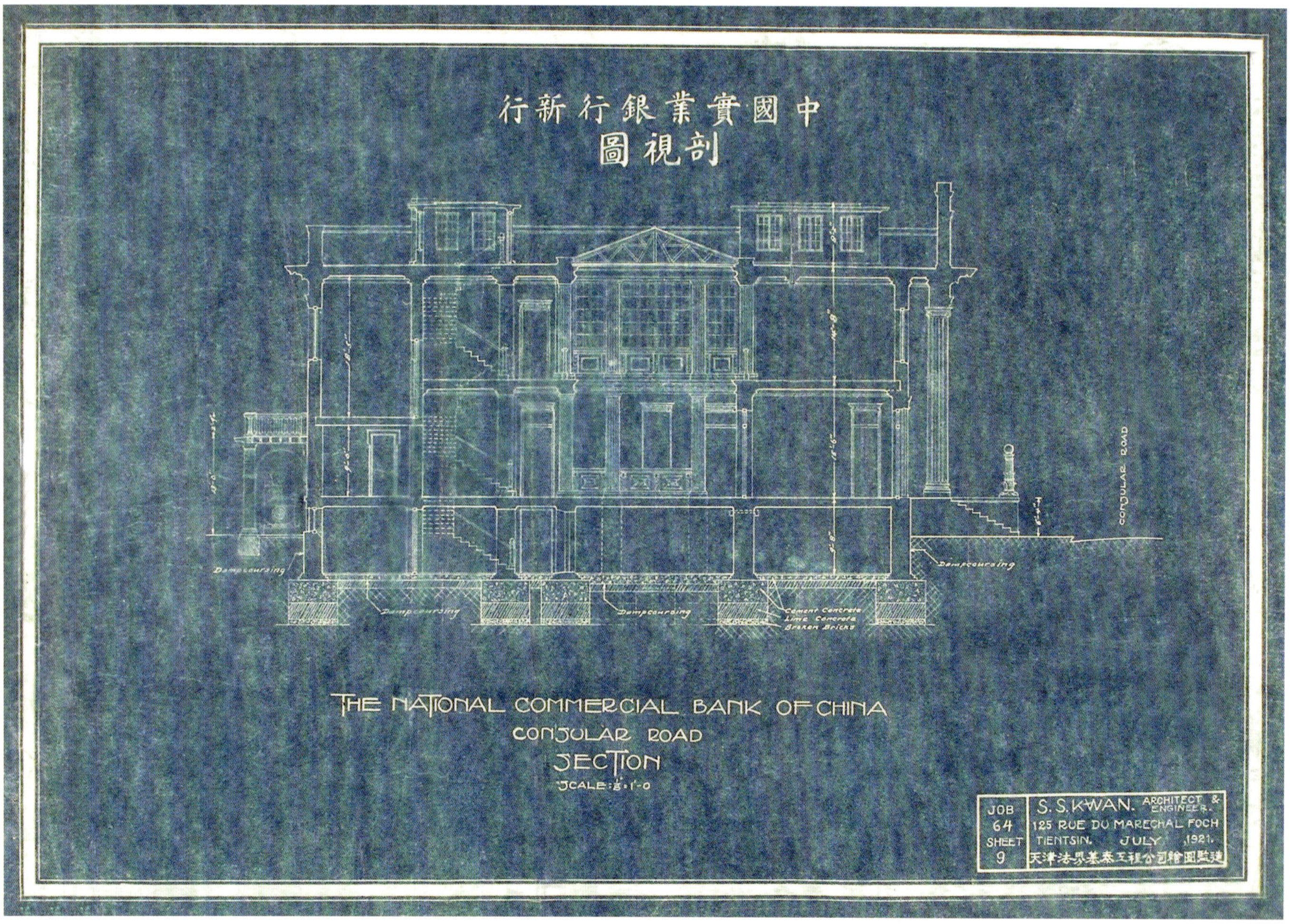

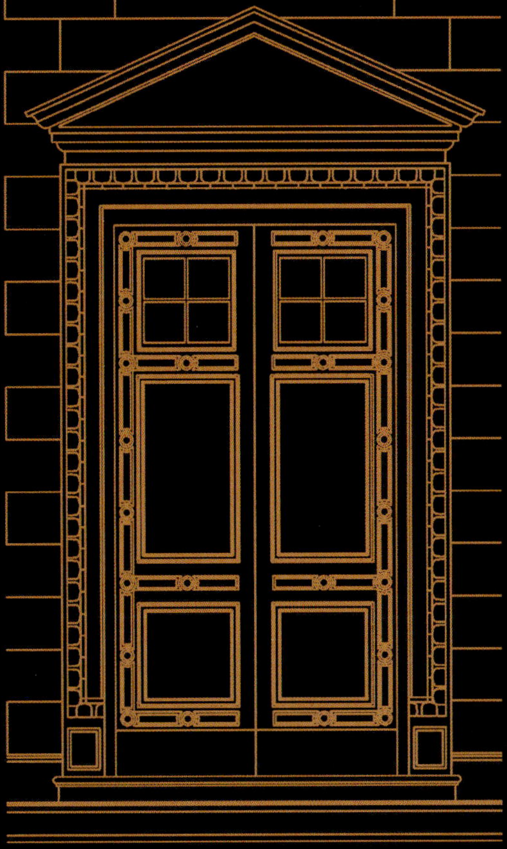

Former French-Chinese Bank of Commerce and Industry
原中法工商银行

中法工商银行于1923年由原中法实业银行改组而成，为中法合资银行，在上海、天津、北京等地设有分行。天津分行1925年开业，1948年停业。该建筑位于和平区解放北路74~78 号，天津市文物保护单位，特殊保护等级历史风貌建筑，现为办公用房。建筑始建于1919年，1932－1933年、1936年增建，四层混和结构楼房（设有地下室），局部五层。占地面积约1 600 m^2，建筑面积约5 000 m^2。建筑主入口设于街角，两侧通廊由沿弧线对称布置的十棵科林斯巨柱构成，气势宏伟，造型典雅别致，是典型的古典主义风格。

The Former French-Chinese Bank of Commerce and Industry was reorganized from the Banque Industrielle de Chine in 1923. It was a joint-venture bank and had run its branches in Shanghai, Beijing and Tianjin as well. The Tianjin Branch opened in 1925 and ended its business in 1948.

The building formerly serving as the Bank's site sits at 74~78 North Jiefang Road, Heping District, as one of the heritage site under city protection and the special protection historic architecture in Tianjin. This building was initially built in 1919 and expanded in 1932-1933 and 1936 respectively. Now, the building is used as office building, covering an area of about 1,600 square meters, and about 5,000 square meters in floor area. The composite structure is mainly four-storied but five-storied partially, plus its basement. The main entrance is designed at the corner with ten massive Corinthian columns standing symmetrically along its sides, forming the open colonnade. The whole building looks significant and elegant, presenting with the Classical atmosphere.

壁柱柱式 Pilaster

外檐细部 Details of the outer-eaves

侧入口 Side-entrance

外檐细部 Details of the outer-eaves

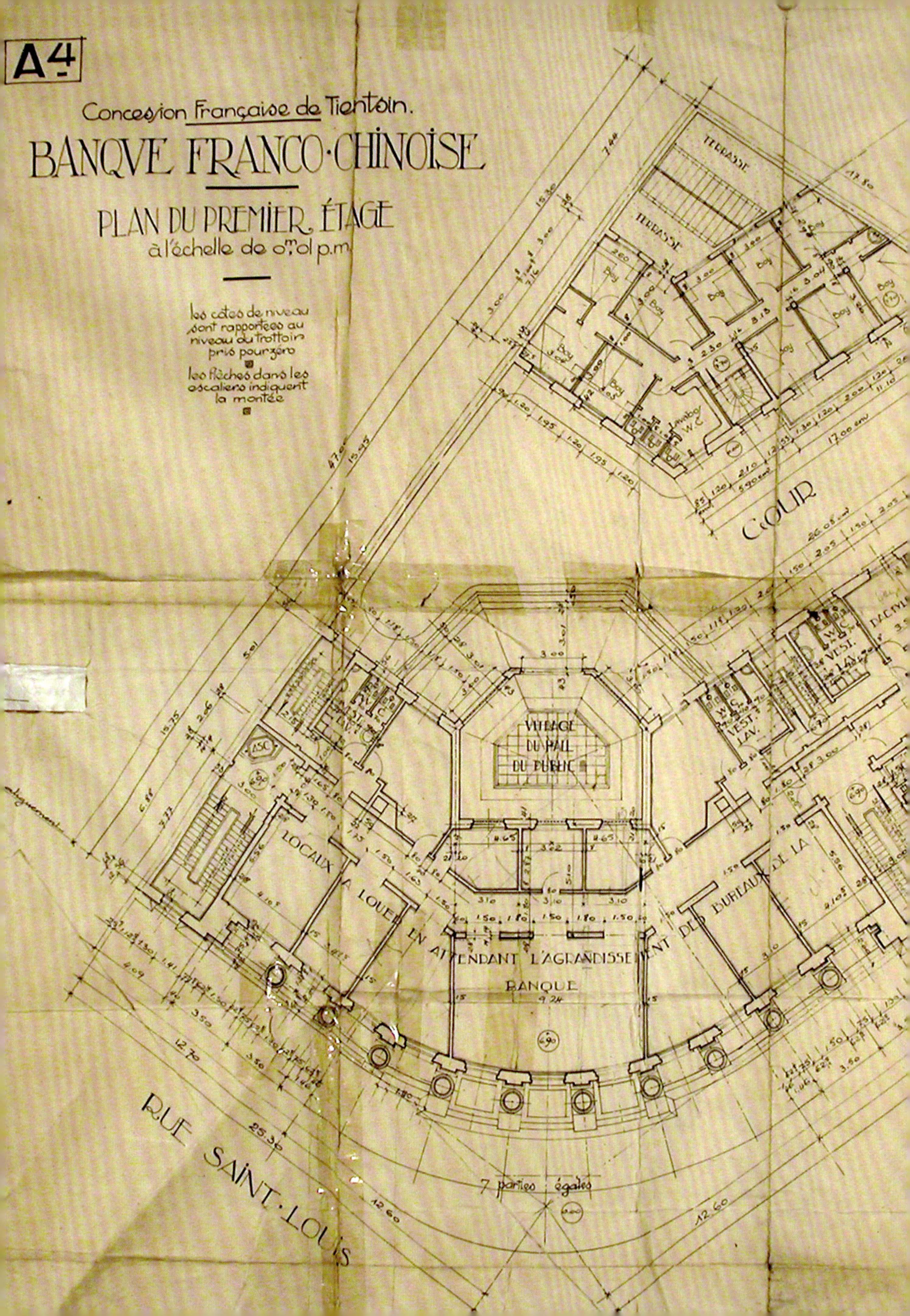
A4
Concession Française de Tientsin.
BANQUE FRANCO-CHINOISE
PLAN DU PREMIER ÉTAGE
à l'échelle de 0m01 p.m.
les côtes de niveau sont rapportées au niveau du trottoir pris pour zéro
les flèches dans les escaliers indiquent la montée
TERRASSE
TERRASSE
Boy
W.C.
COUR
VITRAGE DU HALL DU PUBLIC
LOCAUX A LOUER EN ATTENDANT L'AGRANDISSEMENT DES BUREAUX DE LA BANQUE
W.C.
VEST.
LAV.
ASC
RUE SAINT-LOUIS
7 parties égales

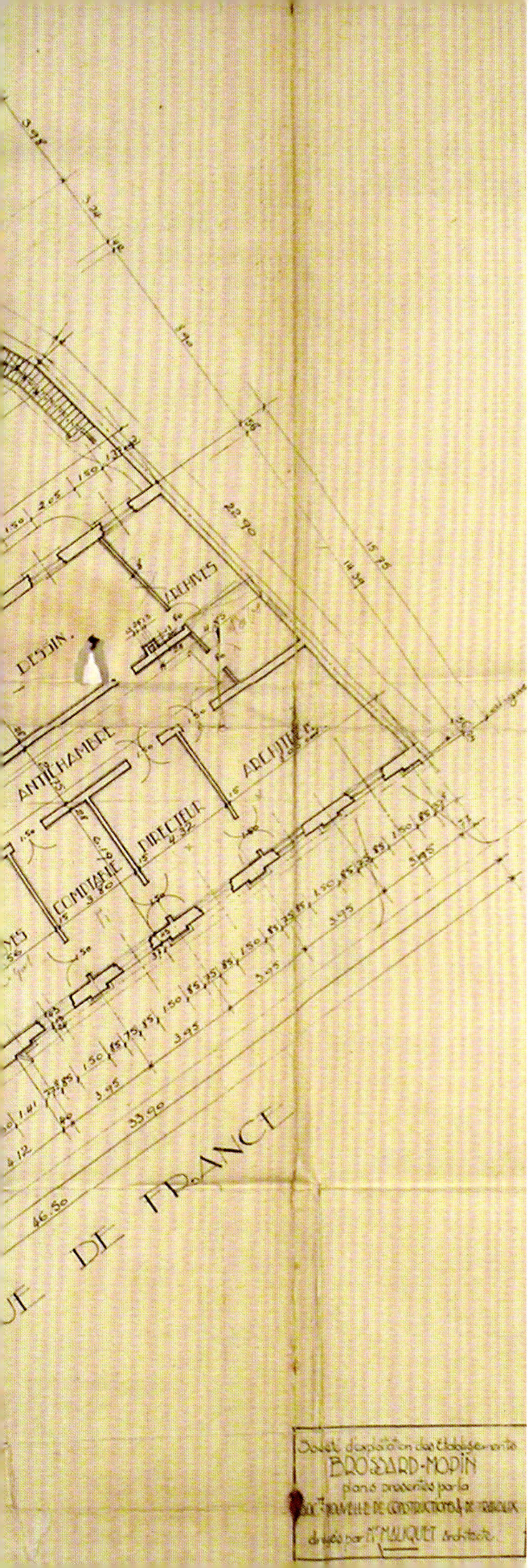

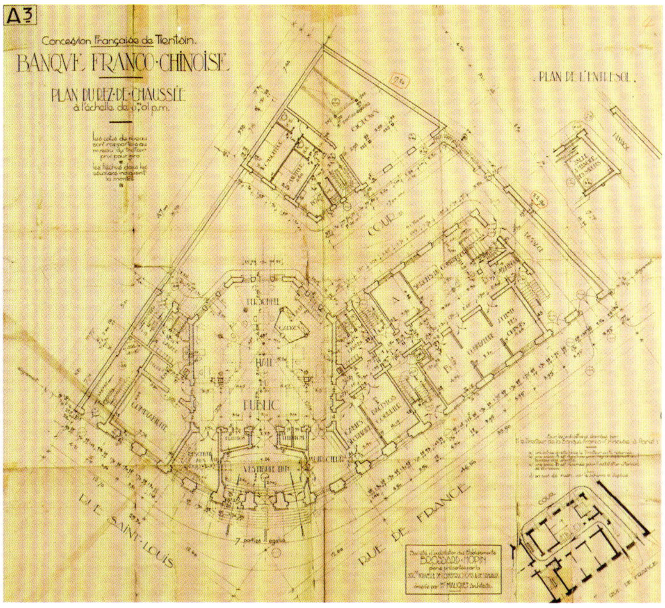

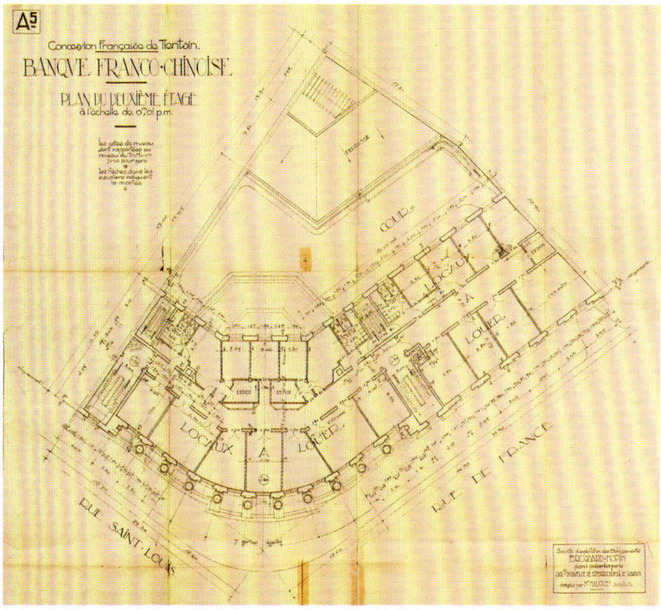

原设计图 Original blueprint

外檐局部 Partial outer-eaves

外景 Exterior view

解放北路

Former Citibank

原花旗银行

原花旗银行位于和平区解放北路90号，天津市文物保护单位，特殊保护等级历史风貌建筑，现为中国农业银行。花旗银行1812年成立于美国纽约，天津分行开业于1916年，独家负责美国对华贸易输出结算，除办理存放款业务外，还发行钞票。是在津美资银行势力最大的一家。1942年太平洋战争爆发后，资产被日本当局没收，抗日战争胜利后重新开业，并依靠政治优势取得天津外商银行的“霸主”地位。1948年12月停止营业。该建筑建于1921年，建筑面积约2 000m^2，由建筑师穆菲和达那设计。混合结构三层楼房，仿希腊古典复兴建筑风格。正入口门前由四棵贯通一、二层的巨大爱奥尼克柱构成开敞柱廊，强化其对称构图。室内大厅内部立有七棵方柱，墙面镶有壁柱，屋顶有装饰雕刻。

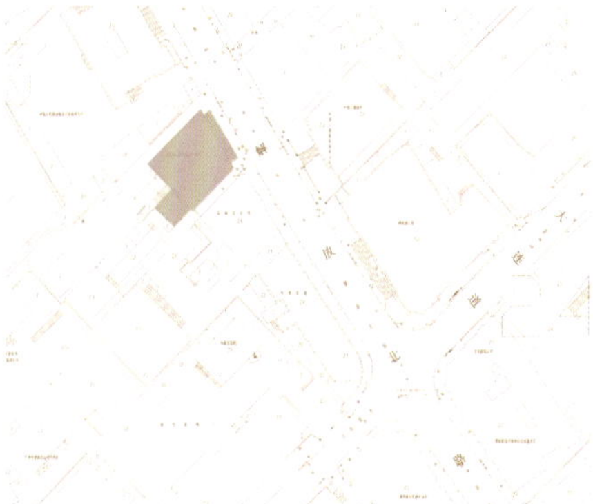

外景 Exterior view

AGRICULTURAL BANK OF CHINA

历史照片 Historic photo

The Former Citibank was situated at 90 North Jiefang Road, Heping District. It is now the heritage site under city protection and the important protection historic architecture in Tianjin and used as the office building of the Agricultural Bank of China. The Citibank was established in New York in 1812 and its Tianjin branch was opened in 1916 and solely responsible for the settlement of America's trade with China, deposits, loans and issuing banknotes. After the outbreak of the Pacific War in 1942, the bank was confiscated by the Japanese invaders. After the 1945 victory, the bank was reopened and gained the place of the dominant foreign bank in Tianjin. In December, 1948, the bank was closed down. This building was designed by MURPHY&DANA and built in 1921. This three-storied building with a floor area of some 2,000 square meters belongs to the classical Revival architecture. The four Ionic pillars standing in symmetry in front of the building formed an open colonnade and the hall was supported by seven square columns. The interior walls were decorated with pilasters and the roof was engraved with decorative patterns.

天花板 Ceiling

Former Yokohama Specie Bank Limited

原横滨正金银行

横滨正金银行创办于1880年，总部设在横滨。天津分行成立于1899年，以经营外汇及存、贷款为主要业务，曾在中国发行钞票。通过贷款、透支，支持日商大量向中国华北输出商品，获得廉价原料，牟取暴利。抗战胜利后，由中国银行接收。

该建筑位于和平区解放北路80号，天津市文物保护单位，特殊保护等级历史风貌建筑，现为中国银行。建筑建于1926年，占地面积约2 700 m^2，建筑面积约5 000 m^2。由英商爱迪克生和达拉斯（同和）工程司设计，三层混合结构楼房。建筑外檐为石材墙面，造型稳重华丽，其正立面的八棵科林斯巨柱构成的开敞柱廊强调了对称构图，是典型的古典主义风格。

The Bank was established in 1880 with its headquarter in Yokohama, Japan. Its Tianjin Branch was founded in 1899. Its business scope was mainly related with foreign exchange, deposit and loan. And the Bank also issued the banknotes in China. By means of loan and overdraft, the Bank supported the Japanese merchants for exporting their goods to North China and obtaining the cheaper raw material from China in order to seek excessive profits. After the Anti-Japanese War, the Bank was taken over by Bank of China.

Located at 80 North Jiefang Road and built in 1926, the building once occupied by the Japanese Bank was designed by the English designing body, Atkinson & Dallas Architects and Civil Engineers Ltd. It is opened as the heritage site under city protection and the special protection historic architecture in Tianjin. Now, the building is used by Bank of China, covering an area of about 2,700 square meters, and about 5,000 square meters in floor area.

The outer-eaves of this three-storied composite structure are decorated with stone facings, looking steady and smart. A grand colonnade presents with 8 massive Corinthian columns standing symmetrically along the facade of the building. It could be considered as a typical Western architecture in the Classical form.

历史照片 Historic photo

外檐局部 Partial outer eaves

外景 Exterior view

中國銀行

主入口门 Main Entrance

外檐局部 Partial outer-eaves

铜门上的花饰 Decoration on the bronze door

营业大厅顶棚 Ceiling of the Business Hall

内檐柱头 Column capital under the inner eaves

门套 Marble overdoor under the inner eaves in the Business Hall

贵宾室 Room for the Honored Guests

贵宾室灯座 Lamp holder in the Room for the Honored Guests

壁炉 Fireplace

内檐彩色玻璃门 Stained-glass door under the inner-eaves

二层走廊 Corridor of second floor

↓楼梯 Stairs

Former Korean Bank
原朝鲜银行

原朝鲜银行位于和平区解放北路97~101号，天津市区、县文物保护单位，特殊保护等级历史风貌建筑，现为办公、居住用房混合使用。朝鲜银行由日本投资，1911年在朝鲜京城建立总行，天津分行于1918年开业。抗日战争胜利后，由中国政府按敌伪产业予以接收。该建筑建于1918年，占地面积约1 100 m^2，建筑面积约3 500 m^2。混合结构三层楼房。为仿希腊古典复兴建筑风格，但外檐墙面、柱式均采用清水红砖，体现了地方材料和鲜明的个性，色彩明快温馨；主入口设于转角处，建筑体量处理简洁明快，细部处理精致协调。

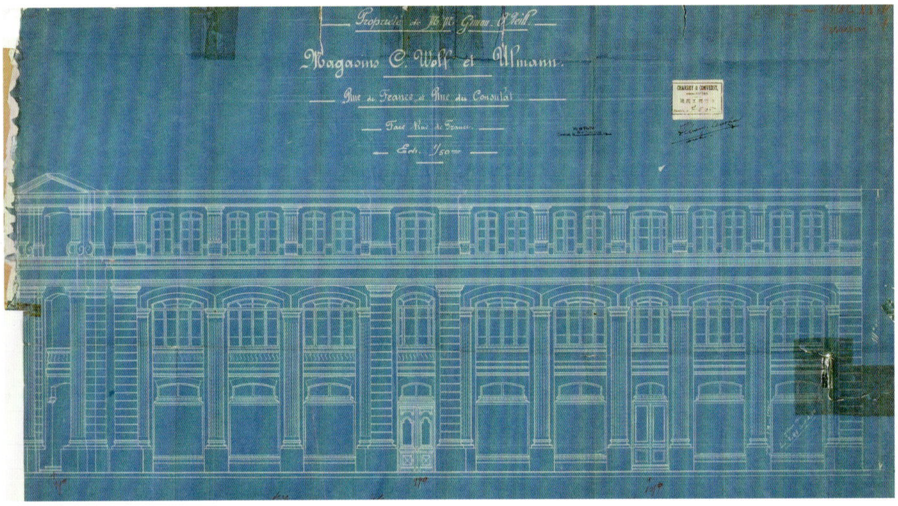

原设计图 Original blueprint

The Former Korean Bank, situated at 97~101 North Jiefang Road, Heping District, is now the heritage site under city/county protection and the special protection historic architecture in Tianjin and is used as an office and residential building. It was a Japanese-funded bank established in the capital of Korea in 1911. The Tianjin branch was opened in 1918. After the victory of the Anti-Japanese War, the Chinese government confiscated the bank as Japanese property. This building was built in 1918, covering an area of some 1,100 square meters and producing a floor area of about 3,500 square meters. It is a three-storied building of the classical Revival style. The red brick exterior walls and pillars exhibit unique local charming. It is a perfect combination of a simple framework and exquisite details.

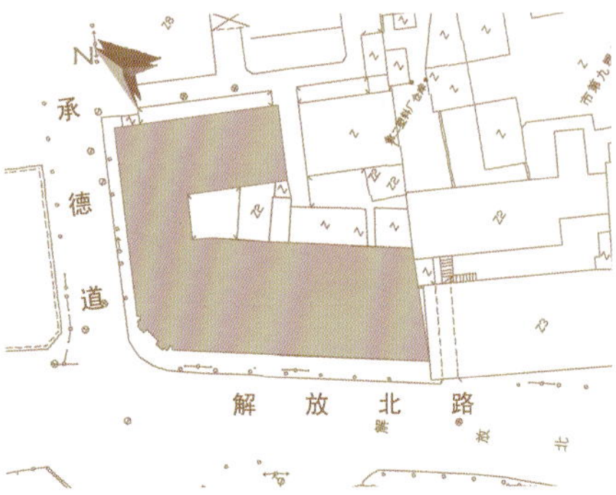

外檐局部 Partial outer-eaves

↓主入口外景 Exterior view of the Main Entrance

↓主入口 Main Entrance

外景 Exterior view

Former Banque Belge Pour L'Etronger

原华比银行

原华比银行位于和平区解放北路104号，天津市区、县文物保护单位，重点保护等级历史风貌建筑，现为中国建设银行。华比银行属比利时通用银行系统，创立于1902年，总行设在布鲁塞尔，天津分行于1906年开业，主要经营存款、外汇及抵押放款并发行钞票。该建筑于1922年建成，华比银行天津分行同年搬入。除银行自用外，建筑的二层还对外出租，比利时领事馆、美商德士古股份有限公司等均曾租用为办公用房。1956年华比银行天津分行结业。该建筑由比商义品公司设计，占地面积约2 400 m²，建筑面积约3 000 m²。混合结构三层楼房，外檐主立面为石材饰面，立面简洁明快，为当时简约的现代主义建筑风格。室内功能设计得当，装饰考究，木地板、木楼梯等设施保存完好。

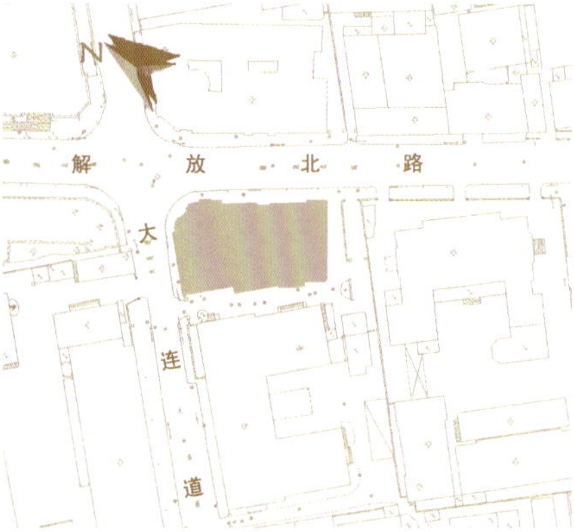

原设计图 Original blueprint

正立面外景 Exterior facade

中国建设银行
China Construction Bank
机动车
禁止驶入
机动车
禁止驶入

外景 Exterior view

The Former Banque Belge Pour L'etronger was situated at 104 North Jiefang Road in Heping District. It is now the heritage site under city/county protection and the important protection historic architecture in Tianjin and used as the office building of China Construction Bank. The Banque Belge Pour L'etronger, a member of the Belgium General Bank system, was established in 1902 and headquartered in Brussels. Its Tianjin branch was opened in 1906 and specialized in deposits, foreign exchange, secured loan and issuing banknotes. Designed by the Credit Foncier d' Extreme Orient and built in 1922, the building covers an area of some 2,400 square meters and has a floor area of about 3,000 square meters. The Tianjin Branch of the Banque Belge Pour L'etronger moved in the same year. The Belgium Consulate, American Texaco Limited Company, etc. used to rent offices on the second floor. In 1956, the Tianjin Branch of Banque Belge Pour L'etronger was closed down. The rooms are well-designed and beautifully decorated. Up to now, the wood floor and the flights are in good condition. The plain stone exterior facade exhibits a simple style of modernism.

壁炉 Fireplace

走廊 corridor

一层门厅 Lobby of the first floor

Former Russo-Chinese Bank
原华俄道胜银行

华俄道胜银行由法国和俄国银行业发起成立，总行设在彼得堡。天津分行1896年开业，以发行的卢布为主要交易筹码，进行投机买卖，代办中国各种税款，1926年停业。该建筑位于和平区解放北路121号，天津市文物保护单位，特殊保护等级历史风貌建筑，现为中国人民银行。建筑建于1900 年，占地面积约700 m^2，建筑面积约2 900 m^2。二层砖木结构楼房。建筑外檐墙面饰以黄色面砖，券形窗口、饰有人字形山花的平窗、弧形转角及盔顶，是一座具有浓郁的俄罗斯风格的古典主义建筑，室内装修讲究，大量精美木饰保存完好。

The bank was originally founded by means of the joint Franco-Russian capital with its headquarter in Petersburg, Russia. It opened its Tianjin Branch the next year in 1896, mainly dealing with the Chinese tax affairs and issuing Rubles to aid its speculative transactions in China. It was closed in 1926.

Originally built in 1910, the former business building of the Branch is located at 121 North Jiefang Road, Heping District. It has been regarded as the heritage site under city protection and the special protection historic architecture in Tianjin. Now, the building is used by The People's Bank of China, covering an area of about 700 square meters, and about 2,900 square meters in floor area.

This two-storied brick-wood structure was designed classically in a typical Russian style, yellow-faced walls, arched windows accompanied with the pedimental windows and curving corners along with huge helmet-shaped roof. The interior decorations have been regarded to be the most elegant and a lot of wooden works have been preserved quite well till now.

历史照片 Historic photo

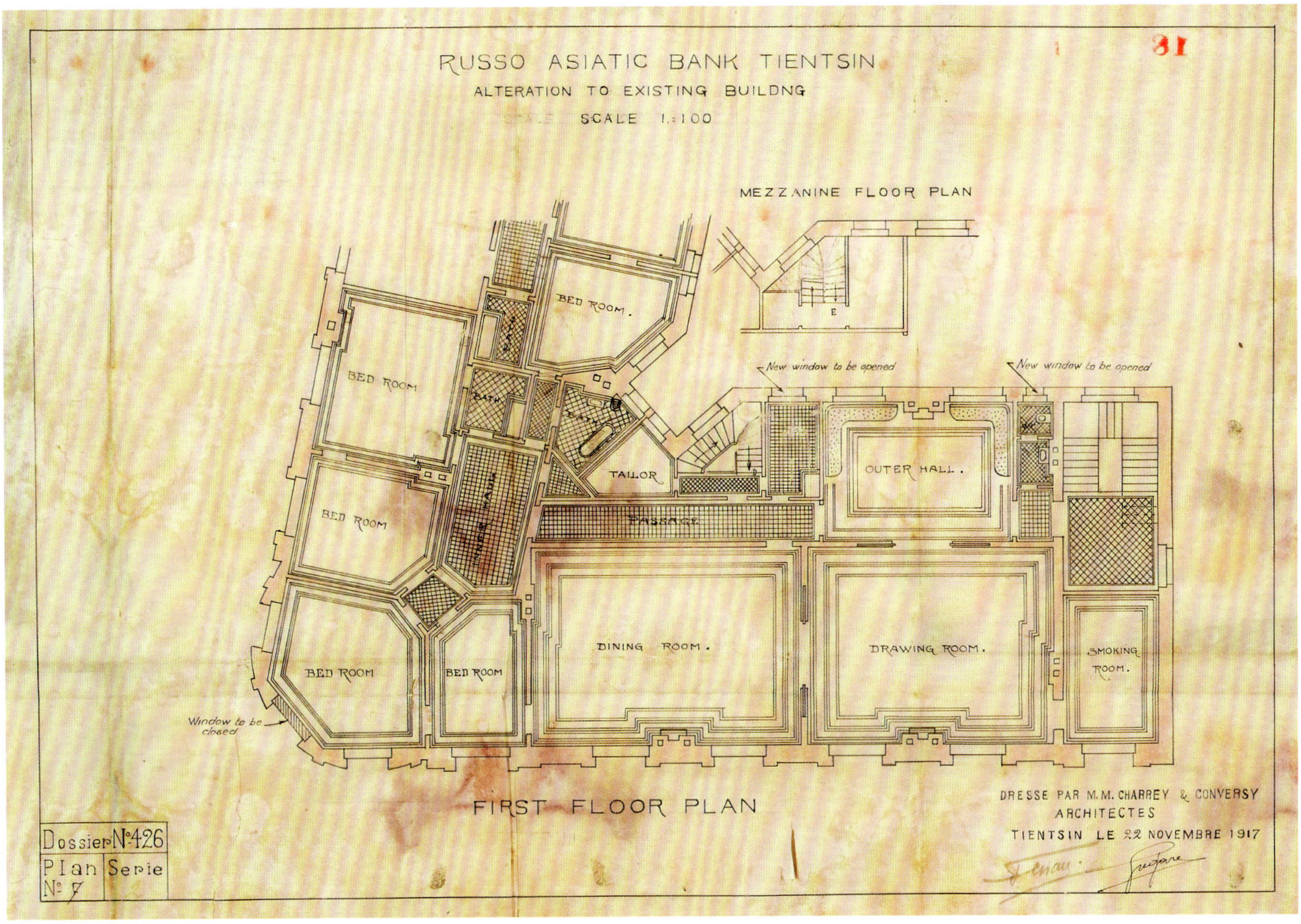

1917年建筑改造设计图 Modification blueprint in 1917

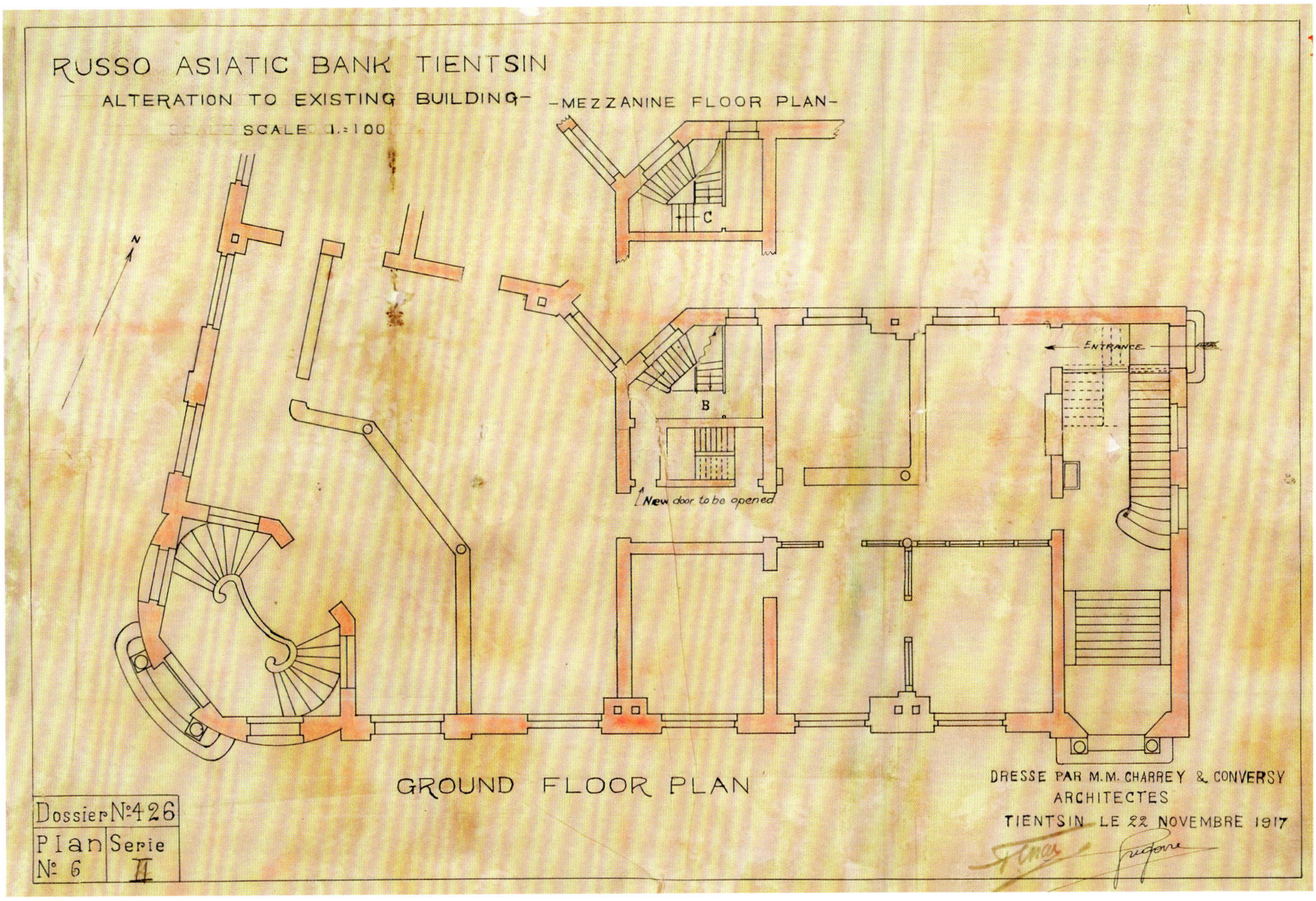

塔楼 Roof tower (Helmet-shaped roof)

↓外檐窗 Windows below the outer-eaves

塔楼内木结构 Wooden structures inside the roof tower

外景 Exterior view

侧入口 Side Entrance

Commercial Buildings

商贸建筑

Astor House Hotel
利顺德大饭店

利顺德大饭店位于和平区解放北路199号，全国重点文物保护单位，特殊保护等级历史风貌建筑，创建于1863年，是中国近代首家外商开办的大饭店。孙中山、美国前总统胡佛等曾在此下榻。该饭店初为平房，1884年扩建为英国古典式三层楼房，是当时英租界最大最高的建筑。1929年增建四层楼房，经多次改造，现两座建筑保存完好，占地面积约3 900 m^2，建筑面积约10 500 m^2。位于街角的三层楼房，建筑造型稳重大方，舒展挺拔，转角处方形角楼强化了竖向构图，外墙装饰丰富且手法简洁，运用了许多古典建筑符号。

历史照片 Historic photo

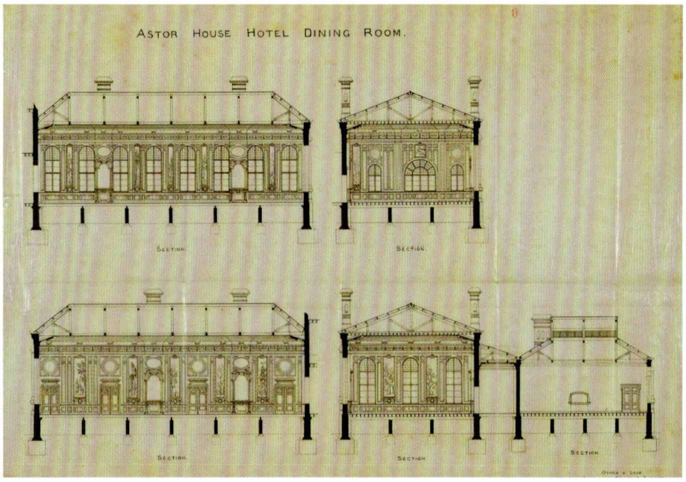

原设计图 Original blueprint

Regarded as the important heritage site under state protection and the special protection historic architecture in Tianjin, Astor House Hotel is located at 199 North Jiefang Road, Heping District, covering an area of about 3,900 square meters, and about 10,500 square meters in floor area. Many great figures have once stayed at the hotel, such as Dr. Sun Yat-sen and Hoover, deceased former president of U.S.

Being the first large-scale hotel operated by the foreigners in the contemporary China, the hotel was founded in 1863 in bungalow-style and expanded to be three-storied hotel in English classical-style in 1884 which was the tallest building in the British concession at that time. Once more in 1929, the Hotel was reconstructed to be four-storied. Both of the buildings have been preserved very well.

The three-storied building at the street corner tastes massive and decent. Its quadrate turret at the turning point was aimed at enhancing visually the vertical sense of the structure. The walls were finished richly but simple. A lot of classical architectural signs were adapted to the designing of the building.

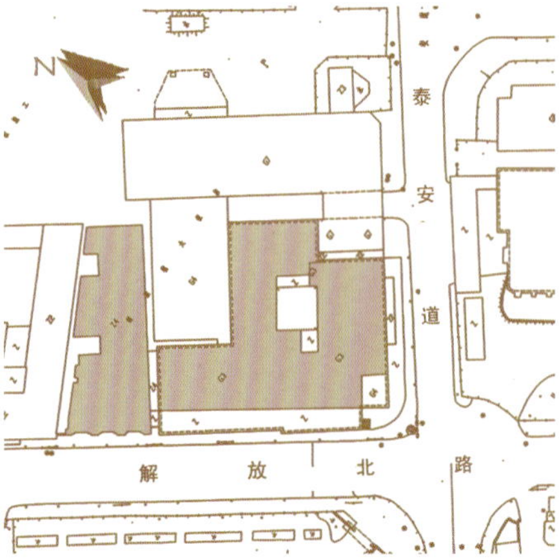

主入口外景 Exterior view of the Main Entrance

↓外檐细部 Details of the outer-eaves

外景 Exterior view

一楼大厅 Lobby on first floor

壁炉 Fireplace

孙中山曾住过的客房 The guest-room at which Dr. Sun Yat-sen once stayed

西餐厅 Western dining-hall

胡佛曾住过的客房
The guest-room at which Hoover, deceased former President of U.S. once stayed

梅兰芳曾住客房 The guest-room at which Mr. Mei Lanfang once stayed

老楼梯 Old-fashioned stairs

老电梯 Old-fashioned elevator

Former Talati House Hotel
原泰莱饭店

原泰莱饭店位于和平区解放北路158号，重点保护等级历史风貌建筑，现为天津第一饭店。饭店创办人泰莱悌为印度流浪汉，原籍孟买，二十世纪初来津，初开设小卖部，后开设永昌泰洋行，富有后加入英国籍转向房地产业。1929年泰莱悌与英国人莱德劳合股筹建泰莱饭店，后泰莱悌设法赶走合伙人，独占了该饭店。1936年，泰莱悌取得了饭店北侧的土地，并增建了六层楼房，与既有建筑连成一体，成为当时规模较大的综合性饭店。占地面积约1 800m^2，建筑面积约7 600m^2。建筑由比商义品公司设计，钢筋混凝土框架结构。首层为商业用房，二层设写字间，三至六层为公寓，每套公寓均设有会客室、餐室、厨房、卫生间，高档公寓还设有备餐室和佣人房间。外檐立面设计简洁，大面积粘贴黄褐色麻面面砖，相间布置水刷石方壁柱，局部装饰简约古典纹样。建筑整体感强，造型典雅，色彩和谐，具有典型的现代建筑特征。

外檐转角细部 Details of the facade at the turning-corner

外檐阳台 Balcony under the outer-eaves

外景 Exterior view

The Former Talati House Hotel, now Tianjin First Hotel was situated at 158 North Jiefang Road, Heping District and it is now an important protection historic architecture in Tianjin. S.B.TALATI, the founder of the hotel, used to be an Indian tramp in Bombay. In the early 1900s, he came to China and first started running a small shop and then Talati Estates Limited. After he earned his first barrel of gold, Talati managed to become a British citizen and set his foot in real estate. In 1929, Talati and Laidlaw, a British joined hands to build Talati House Hotel. But later Talati drove his partner away and made the hotel his own. In 1936, Talati built a six-storied building to the north of the hotel and joined the two buildings to a larger hotel which covers an area of some 1,800 square meters and has a floor area of about 7,600 square meters. This reinforced concrete frame building was designed by the Credit Foncier d'Extreme Orient. The first two floors were used for business while the top four were apartments of which each had its separate living room, dining room, kitchen and toilet. Those luxury ones even had servers and maid rooms. With most of its facade covered with tawny pitting surface front bricks alternating with square granitic plasters, and plain classic patterns as decoration, the Talati House Hotel looks elegant and its harmonious colors and sense of unity emphasize the characteristics of modern architecture.

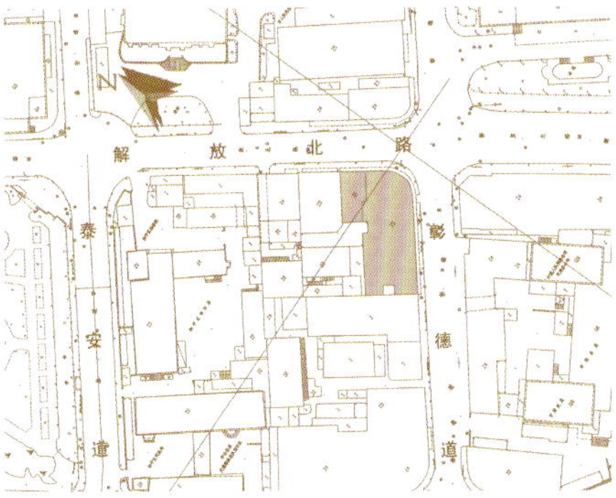

老电梯 Old-fashioned elevator

老电梯控制面板
Control panel of the old-fashioned elevator

外景 Exterior view

一飯店
解放北路

夜景 Night view

Former Huizhong Hotel
原惠中饭店

原惠中饭店位于和平区滨江道141~153号、华中路2号、和平路292号，重点保护等级历史风貌建筑，现为商业用房。该建筑建于1930年，1931年开业，由上海华中营造公司驻津工程部设计，占地面积约2 100 m^2，建筑面积约11 000 m^2，是当时天津著名的饭店之一。为适应当时租界社会不同层次的娱乐与休闲需求，饭店将客房分为优、良、特、福、禄、寿、喜、财八个等级，并设有中西餐部、舞厅、露天影院、露天球场等，出入者多为军阀政客、工商业人士、社会白领等。著名作家曹禺曾以其为社会生活背景，创作出话剧《日出》。该建筑为混合结构建筑，主体五层，局部六层，中部更在六层之上设高耸的三层塔楼，典型的折中主义建筑形式。与周围的劝业场、浙江兴业银行、交通饭店等建筑一起构成了劝业场地区的绝佳建筑空间组合。

Located at 141~153 Binjiang Avenue, 2 Huazhong Road and 292 Heping Road of Heping District, the Former Huizhong Hotel is now an important protection historic architecture in Tianjin and is used for business. Designed by the Tianjin Branch of Shanghai Huazhong Construction Company, Huizhong Hotel was built in 1930 and came into use in 1931. Covering an area of some 2,100 square meters and a floor area of some 11,000 square meters, it was then one of the most well-known restaurants in Tianjin. To meet different requirements of the concessions in Tianjin, rooms in the hotel were classified into eight types and dance hall, open cinema, Chinese cuisine and Western cuisine were available to the guests, most of whom were warlords, politicians, businessmen and elites. With the story happened in this hotel as the original material, Cao Yu, the great playwright, wrote out Sunrise. The Former Huizhong Hotel was a five-storied building, part of it being six-storied, with a three-storied high tower on the top, belonging to the eclectic architecture. The Quanyechang Mall, National Commercial Bank, Communications Hotel and Huizhong Hotel had made a perfect combination of architectures of different styles.

历史照片 Historic photo

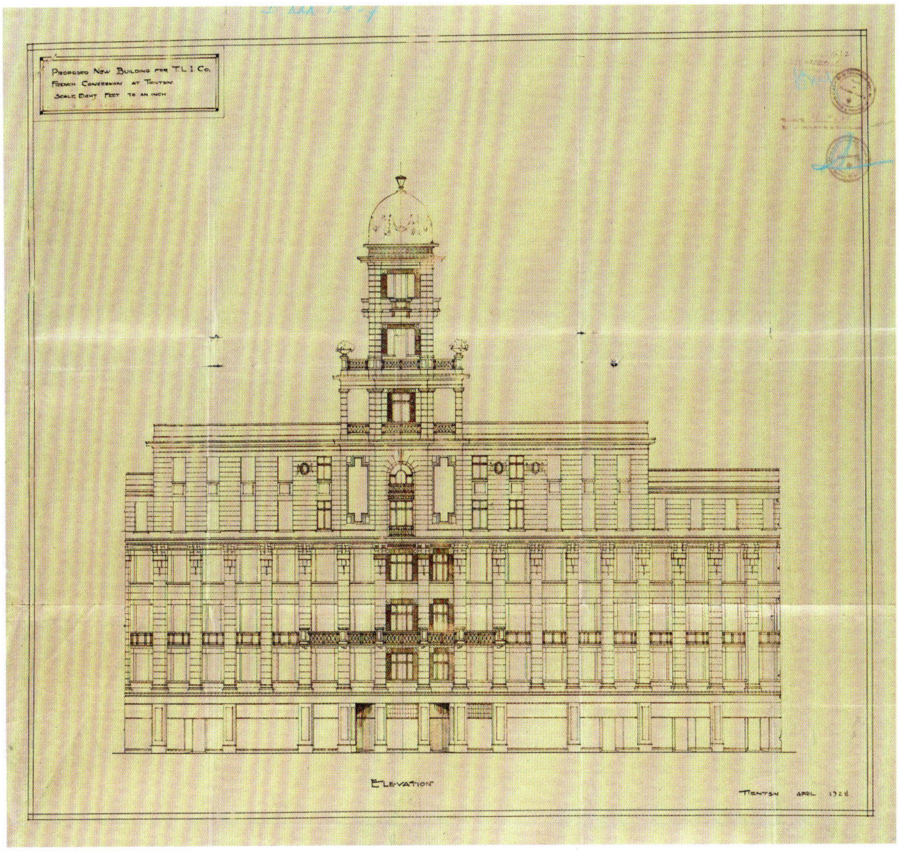

原设计图 Original blueprint

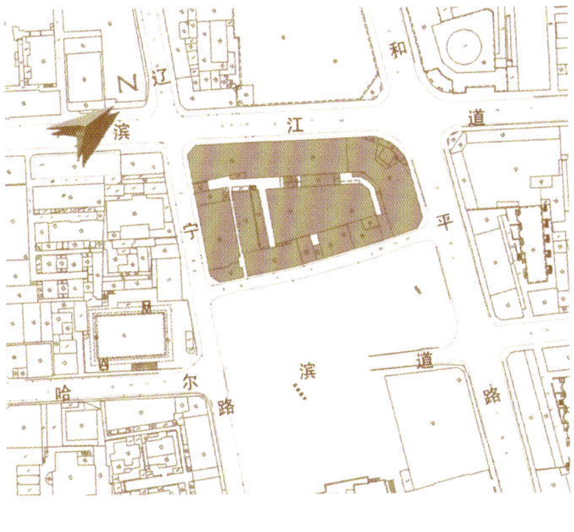

正立面渲染图 Facade Rendering

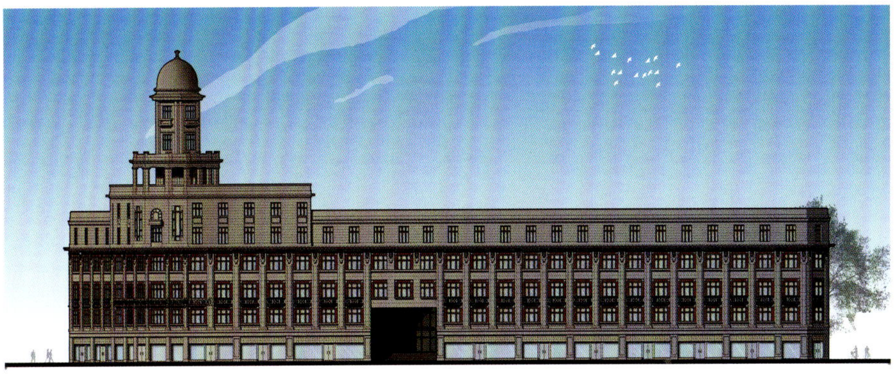

外景 Exterior view

外檐局部 Partial outer-eaves

Former National Grand Hotel

原国民饭店

国民饭店位于和平区赤峰道58号，天津市区、县文物保护单位，重点保护等级历史风貌建筑，为中国商人租用陆安公司土地投资建造，是当时天津豪华饭店之一，现为商业用房。1926年中国共产党在此召开了中华全国铁路总工会第三次代表大会。1934年，爱国将领吉鸿昌在此从事抗日活动时被国民党特务刺伤并逮捕。抗日战争期间，中共天津市委秘密机关和联络站“知识书店”也设于此。该建筑建于1923年，占地面积约3 100m^2，建筑面积约6 400m^2。由乐利工程司设计，原设计四层，实际建造为三层，钢筋混凝土内框架结构，平屋顶，檐部出挑。外檐首层为混水墙面，拱券窗，二层以上为清水砖墙，附壁柱装饰，方窗以山花装饰。建筑南立面采用对称构图，纵横三段式，中部凹进。院落宽敞，内设半球形盔顶凉亭两座，正门为塔什干柱式支撑的门楼，庄重大方。

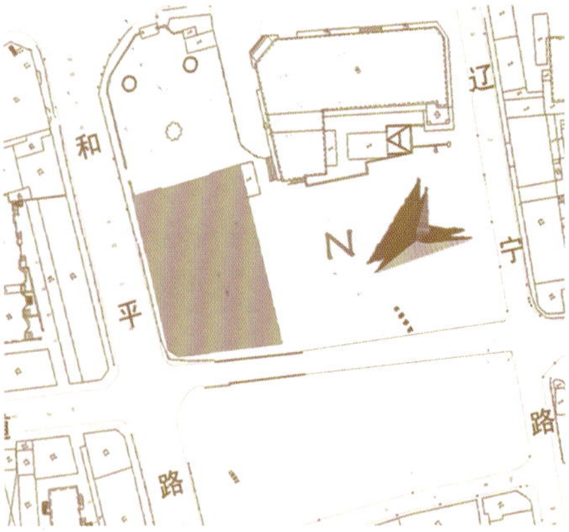

The Former National Grand Hotel was situated at 58 Chifeng Avenue in Heping District. It is the heritage site under city/county protection and the important protection historic architecture in Tianjin and is now used for business. It used to be one of the most luxurious hotels in Tianjin and was built on the land rented from Lu'an Company by a Chinese merchant. In 1926, the Chinese Communist Party held the third Plenary Session of All-China Federation of Railway Workers Unions here and in 1934 Ji Hongchang, the patriotic general, was stabbed and arrested by the agents of KMT in this hotel. During the War of Resistance against Japan, the secret office of Tianjin municipal Party committee and the liaison location—Knowledge Bookstore were both located here. This three-storied hotel with a floor area of about 6,400 square meters was built in 1923, taking up an area of some 3,100 square meters. However, it was originally designed as four-storied by Leli Engineering Company. It is a reinforced concrete inner frame structure construction with a flat roof, overhanging eaves and a gateway supported by Tashkent pillars. Dry walls of red bricks form the decorated exterior of the first floor and plain brick walls are used for the second and third floor. The square windows are decorated with pediments. Two semi-arch arbors are found inside the yard. The building is symmetrical, elegant and spacious.

历史照片 Historic photo

主入口 Main Entrance

历史照片 Historic photo

正立面渲染图 Facade Rendering

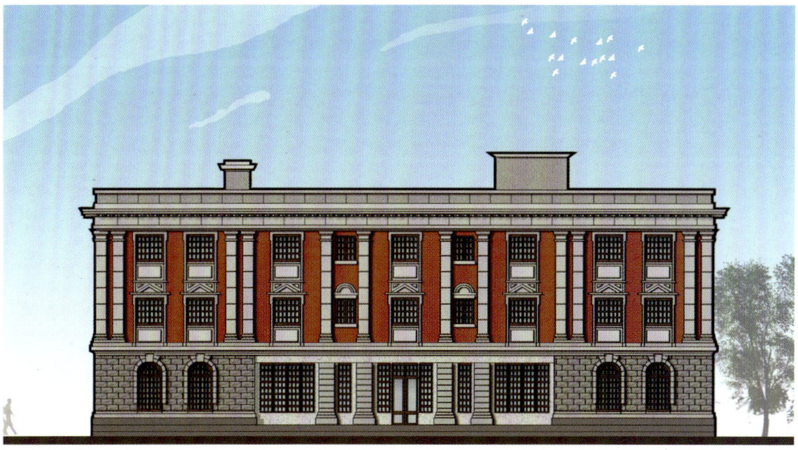

外景 Exterior view

1923

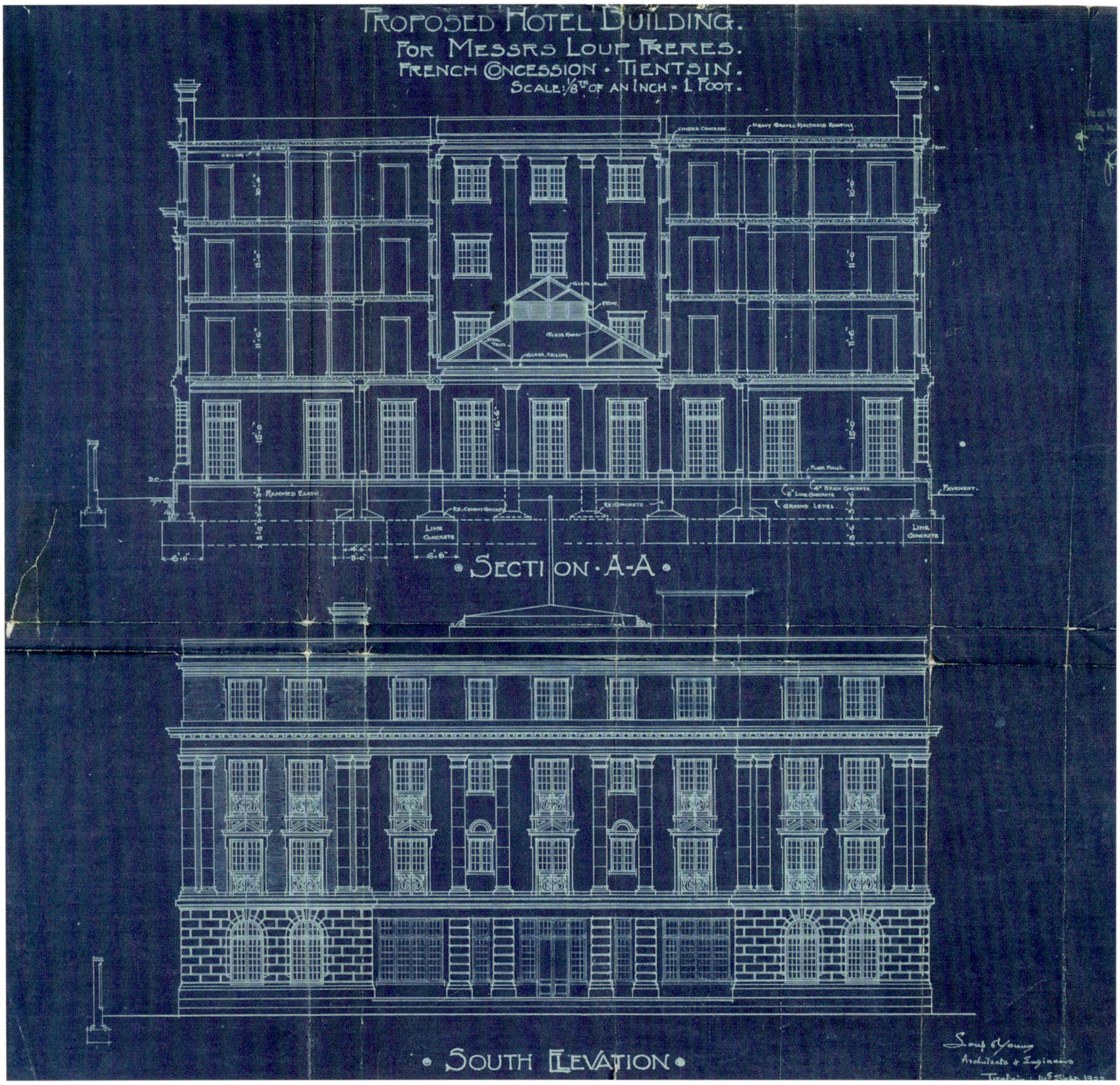

原设计图 Original blueprint

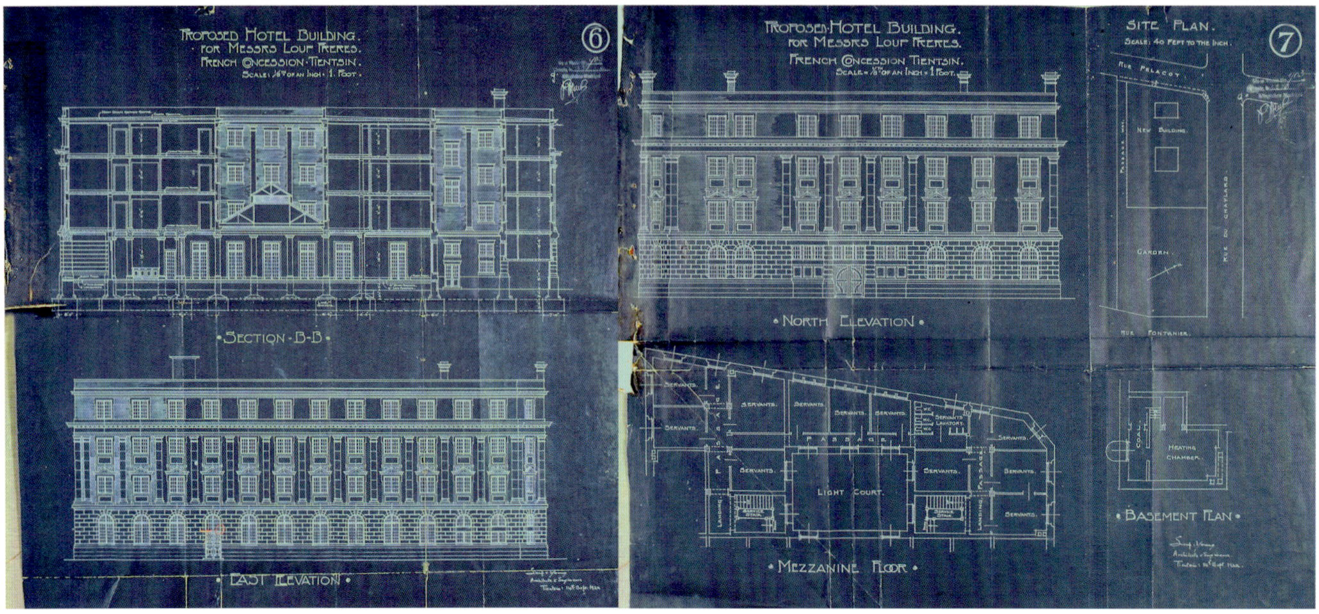

外檐山花 Pediment under the outer-eaves

↓院内凉亭 Summerhouse in the courtyard

Quanyechang Mall

劝业场

劝业场建于1928年，是当时华北地区规模最大的百货商场，“天津劝业场”匾额为著名书法家华世奎所书。场内设有天华景戏院、天宫影院、天乐评戏院等8个娱乐场所，合称“八大天”，集购物、娱乐、休闲于一体，是中华著名商业老字号，并一度成为天津的象征。该建筑位于和平区滨江道152~166号，全国重点文物保护单位，特殊保护等级历史风貌建筑，由法籍工程师穆乐设计，法商永和营造公司建造，占地面积约3 000 m^2，建筑面积约16 500 m^2。建筑为框架结构五层楼房（局部八层），石材饰面，窗户形式多样，建筑形式带有折中主义色彩。

历史照片 Historic photo

华世奎所书匾额 Tablet inscriptions By Mr.Hua Shikui

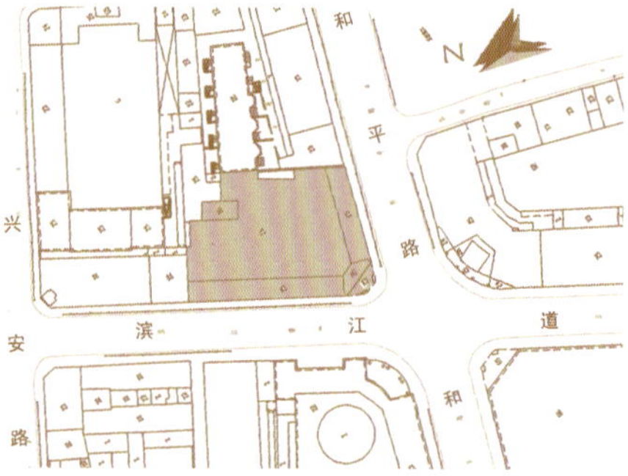

Being the largest scale department store in North China founded in 1928, the Quanyechang Mall was honored with a tablet inscribed by Mr. Hua Shikui, one of the most famous calligraphers in China. Eight grant recreation groups all named with “Tian” got together in the store, including Tian-huajing Theatre, Tian-fu Cinema, and Tain-le Stage for Ping Opera, etc. The Mall could serve the people for their shopping, enjoyment and amusement all within its body. Well-known as one of the outstanding enterprises with long history in China and one of the symbols for the City, it has been classified as the important heritage site under state protection and the special protection historic architecture in Tianjin, covering an area of about 3,000 square meters, and about 16,500 square meters in floor area.

located at 152~166 Binjiang Avenue of Heping District, the building was designed by Mr. P. Muller, the architect from a French engineering company-Hunke & Muller. The building is mainly five-storied and eight-storied partially and has got its walls finished with stone facing. And it has its windows in various forms. The whole architecture tastes to be somewhat an eclectic work considering its designs.

外景 Exterior view

丰富的外檐装饰 Luxuriant designs of the outer-eaves

外檐细部 Details of the outer-eaves

塔楼 Towered roof

入口拱形挑檐 Arched overhang-eaves above the Entrance

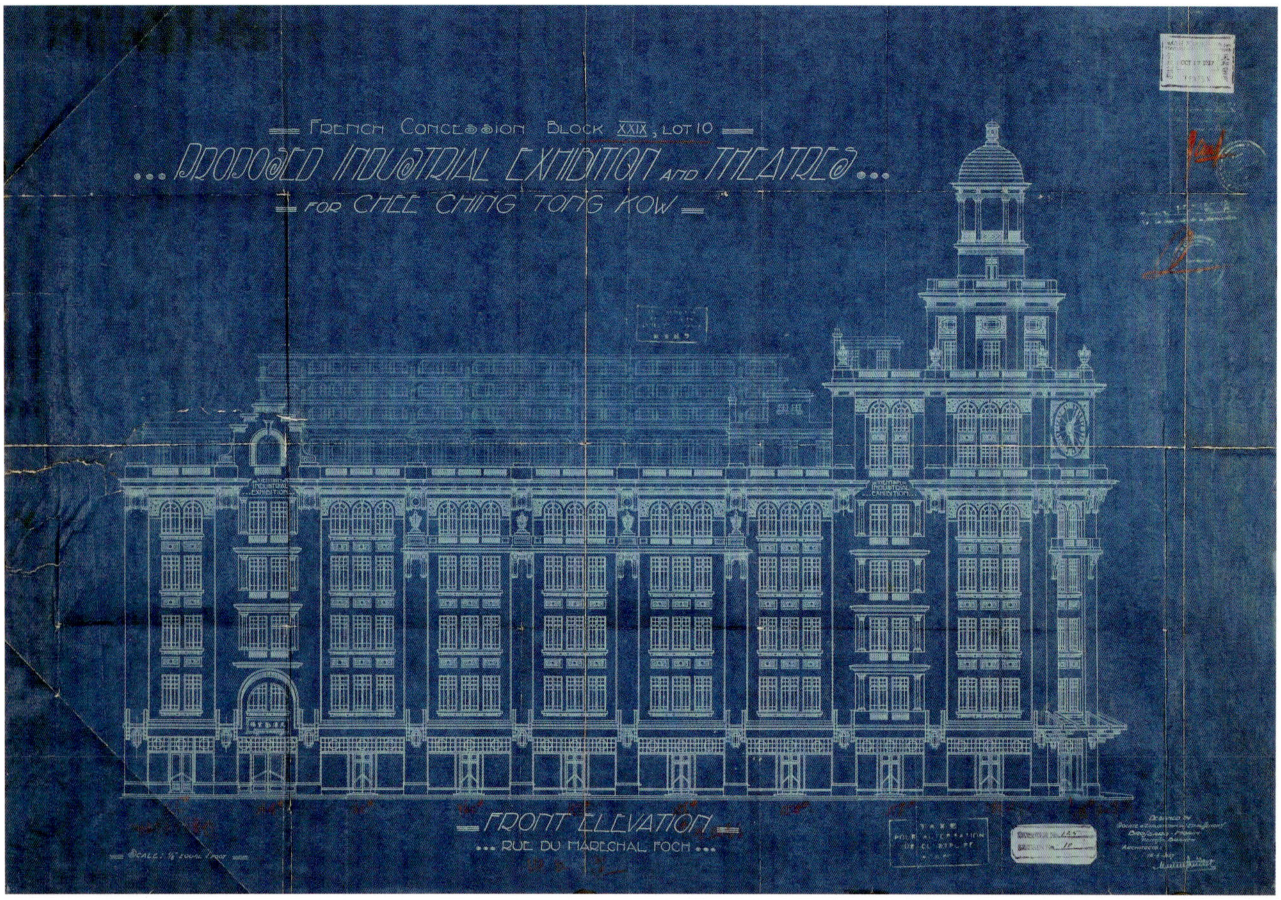

原设计图 Original blueprint

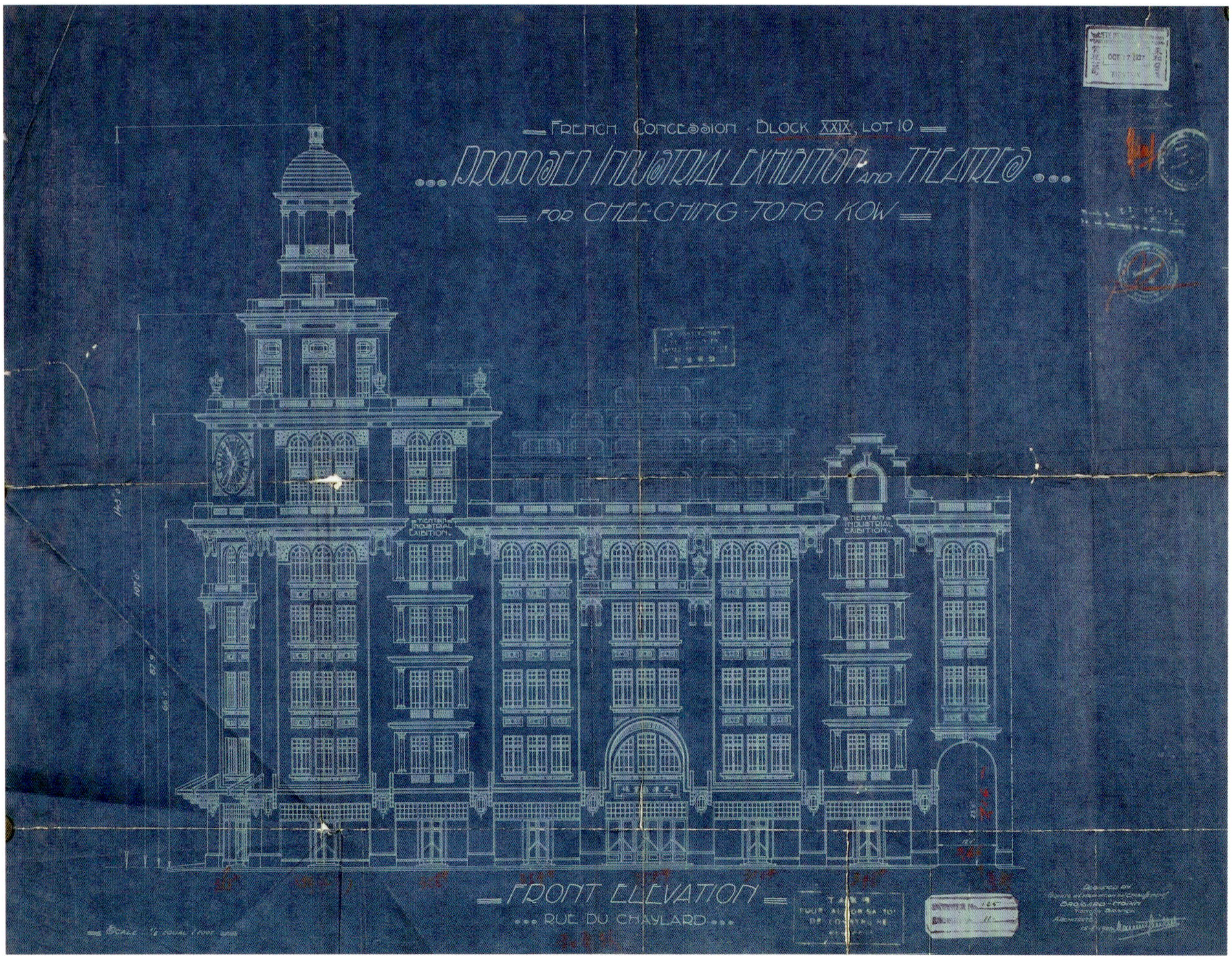

夜景 Night view

Former Laojiuzhang Silk Shop

原老九章绸缎庄

原老九章绸缎庄位于和平区和平路47号，天津市区、县文物保护单位，重点保护等级历史风貌建筑，现为办公用房。老九章绸缎庄为清末一位排行第九的章姓沪商在天津开办，故以“老九章”命名。该店经营绸缎、呢绒，以品种丰富档次齐全享有盛名。该建筑为砖混结构平顶楼房，占地面积约500 m^2，建筑面积约1 900 m^2，原为三层，后增建四层，局部五层。混水墙面，外墙用附壁柱装饰。主入口位于建筑转角处，以柱承托，形成二、三楼阳台，三层檐口有山花。增建部分沿用下面的附壁柱，局部五层在主入口上方做三连拱券窗，屋顶设方形露台。具有古典主义建筑特征。

外檐细部 Details of the outer-eaves

檐口细部 Cornice details

Situated at 47 Heping Road, Heping District, the Former Laojiuzhang Silk Shop is now the heritage site under city/county protection and the important protection historic architecture in Tianjin and is used as an office building. It was set up by a Shanghai businessman who was the ninth child of a Zhang's family in the late Qing Dynasty, hence it was named Laojiuzhang Silk Shop. It sold mainly various silks and woolen cloth. The shop, covering an area of 500 square meters and having a floor area of 1,900 square meters, was a three-storied brick and concrete structure building with a flat top and later a fourth floor was built on the top and even a fifth floor for some parts. The main entrance is at the corner of the building and pilasters are attached to the furred wall. The pillars supporting the main entrance are also used to support the balconies on the second and third floor. Moreover, pediments are used to decorate the cornice and arch windows are designed on the fifth floor and square balcony was built on the roof. It belongs to the Classical architecture style.

外景 Exterior view

外檐转角局部 Details of the facade at the turning-corner

外檐细部 Details of the outer eaves

外景 Exterior view

主入口 Main Entrance

Bo Hai Building

渤海大楼

渤海大楼位于和平区和平路275~281号，天津市文物保护单位，特殊保护等级历史风貌建筑，建于1936年，法商永和营造公司设计，是当时天津最高、最新式的商业、公寓混合大楼，也是当时天津市中心的标志性建筑，现为商业用房。该建筑为钢混框架结构八层楼房（局部十层），高47.47 m，占地面积约500 m^2，建筑面积约4 500 m^2。外墙面粘贴褐色饰面砖，色彩稳重大方。建筑立面强调竖向构图，体量庄重挺拔，属典型的现代主义风格。

历史照片 Historic photo

Bohai Building was designed by the French engineering company—Hunke & Muller and built in 1936. It served both for commerce and resident, regarded as the highest and the modernist building in the city and as symbolic construction of the downtown.

This reinforced concrete structure is mainly eight-storied and partially ten-storied. Its walls were finished with brown tile facing, looking more massive and elegant. Vertical format was adopted for its designing, making the building taste highly-erective visually. It is regarded to be in modern style typically.

Being the heritage site under city protection and the special protection historic architecture in Tianjin, Bohai Building sits at 275~281 Heping Road, Heping District. Now, the building is used as commercial building, covering an area of about 500 square meters, and about 4,500 square meters in floor area.

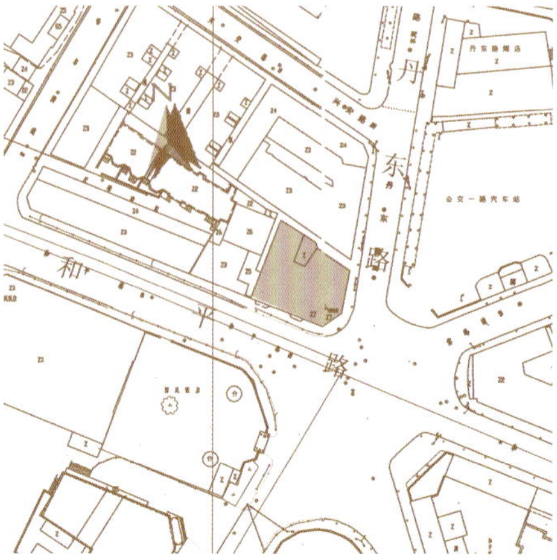

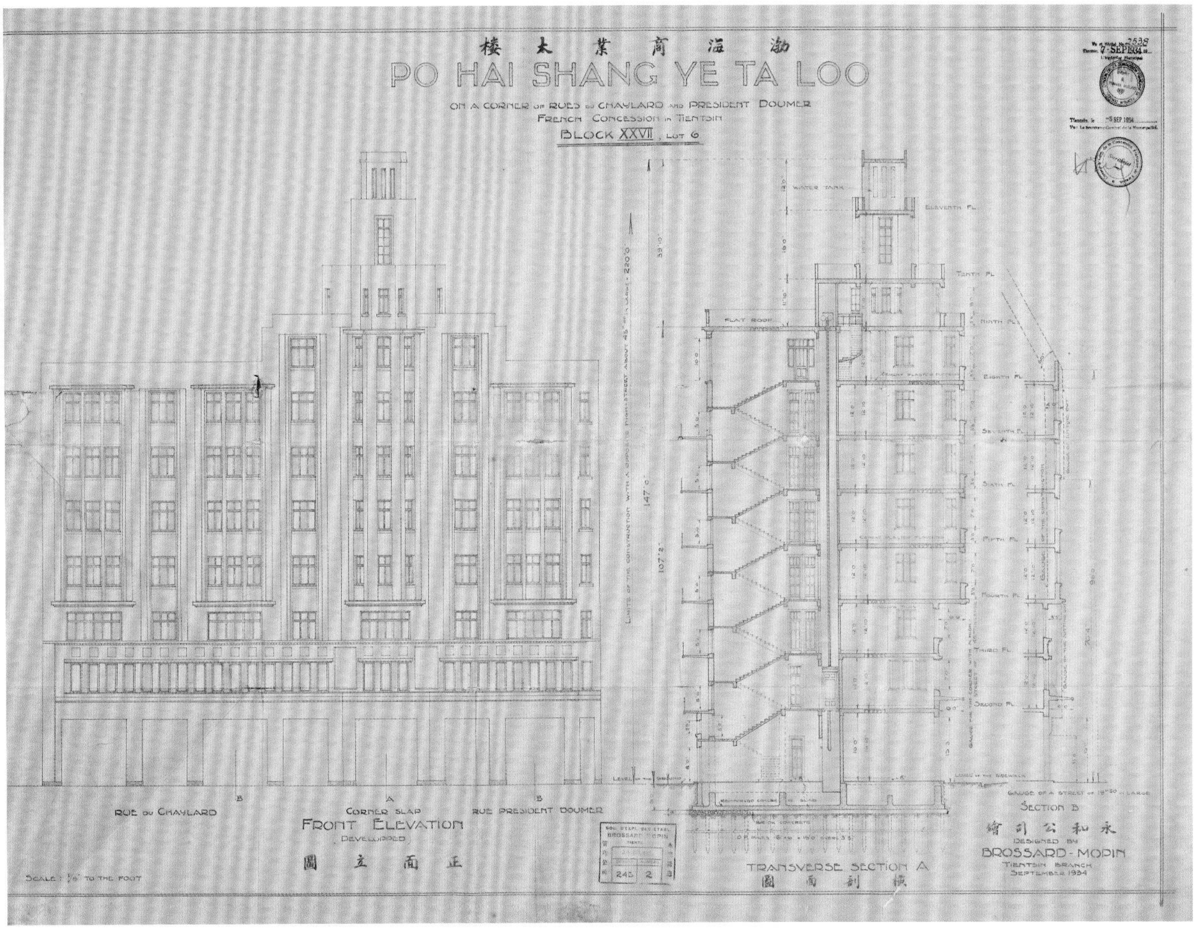

原设计图 Original blueprint

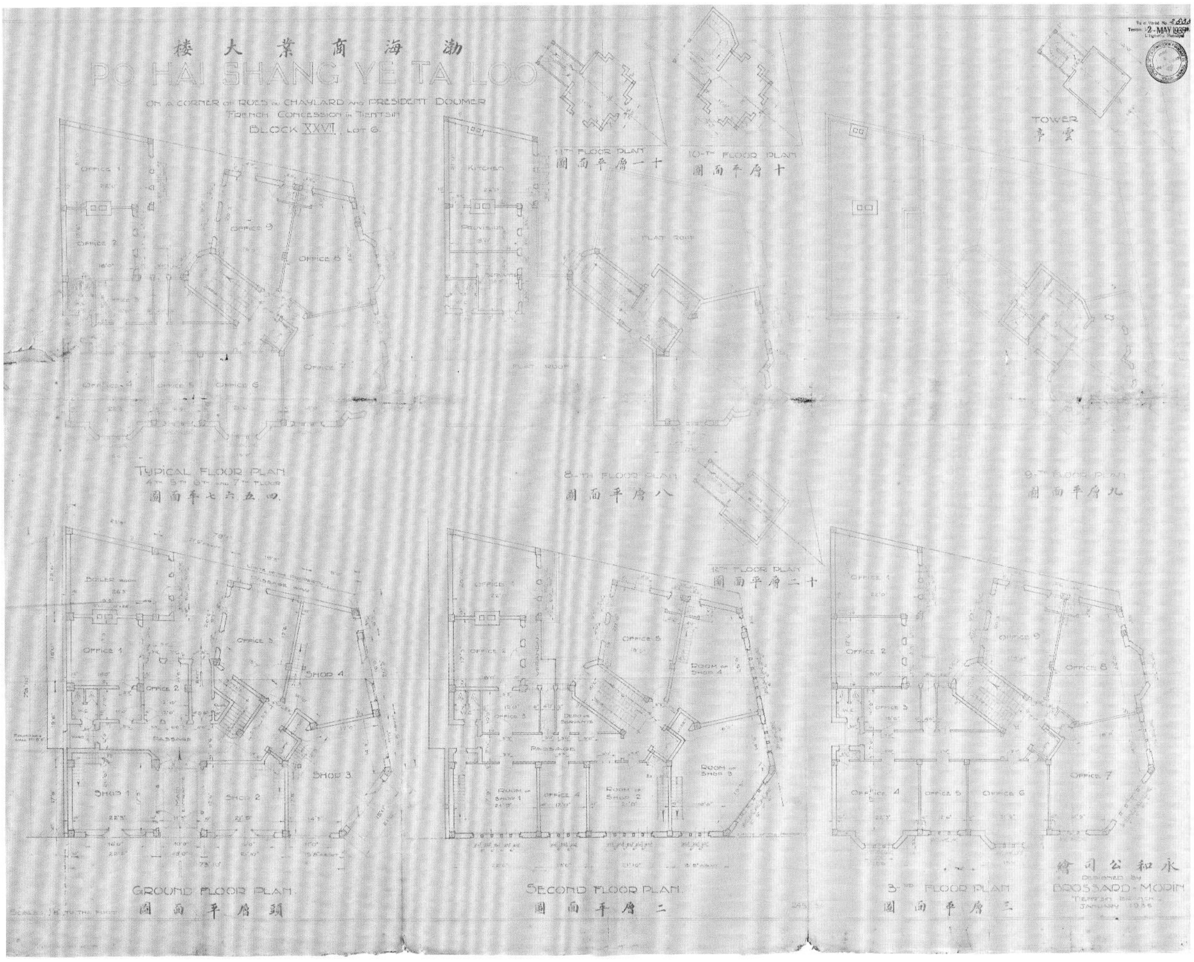

外景 Exterior view

塔楼 Roof tower

Vu et Vérifié No 2538
Tientsin, le 7-SEP 1934 19...
L'Ingénieur Municipal
CONSEIL D'ADON DE LA CONCESSION FRANÇAISE DE TIENTSIN
VOIRIE & TRAVAUX PUBLICS
Tientsin, le -8 SEP 1934
Vu: Le Secrétaire Général de la Municipalit
Conseil d'Adon de la Concession Française de Tientsin
Secrétariat

法租界工部局审批图章
Approval seal of the French Industry Bureau

外檐局部
Partial outer-eaves

Former Watsons

原屈臣氏大药房

原屈臣氏大药房位于和平区解放北路150号，紧邻泰莱饭店，一般保护等级历史风貌建筑，现为办公用房。屈臣氏大药房1886年在天津开业，专营西药，二十世纪初迁至现址。该建筑为四层混合结构，占地面积约600 m^2，建筑面积约2 100 m^2。立面水泥饰面，二至三层有通高爱奥尼克方壁柱，壁柱及窗间墙上有简化的西洋古典建筑装饰图案，简约大方。

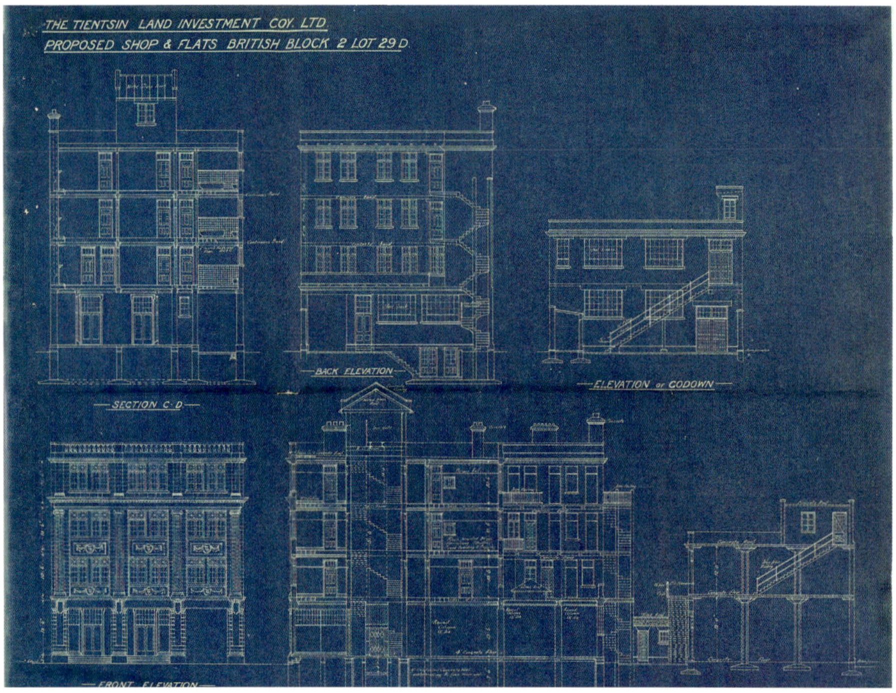

原设计图 Original blueprint

The Former Watsons is at 150 North Jiefang Road, Heping District, next to the Former Talati House Hotel. It is now a general protection historic architecture in Tianjin and is used as an office building. Watsons was opened in Tianjin in 1886 and sold mainly Western medicine and was moved to this building in the early twentieth century. It is a four-storied building which covers an area of 600 square meters and has a floor area of about 2,100 square meters. Cement is applied to the facade and square Ionic pillars are used between the second and the third floor. Simplified Western classic patterns are used to decorate the mural columns, demonstrating a simple style of Western classical architecture.

外景 Exterior view

↓外檐细部 Details of outer-eaves

Former Location of Jardine, Matheson & Co, Ltd.

原怡和洋行大楼

原怡和洋行大楼位于和平区解放北路157号，天津市文物保护单位，特殊保护等级历史风貌建筑，现为办公用房。怡和洋行是最早向中国贩卖鸦片的英国洋行，总行设在上海。天津分行开设于1867年，是天津第一家由外商开办的以从事远洋运输为主的洋行，行址初设紫竹林码头。该行早期以代理船舶为主，1872年将所辖轮船改组为华海轮船公司，1881年在上海成立怡和轮船公司，天津怡和洋行船务部也相应改组为怡和轮船公司天津分公司。该建筑建于1921年，占地面积约1 800 m^2，建筑面积约2 800 m^2，二层混合结构楼房。建筑呈对称布局，主入口位于立面中段，两侧各有一棵科林斯巨柱。建筑体量稳重，装饰简洁，室内装饰豪华大气，具有古典主义建筑特征。

正立面渲染图 Facade Rendering

Standing at 157 North Jiefang Road, the former location of Jardine, Matheson & Co, Ltd. is the heritage site under city protection and the special protection historic architecture in Tianjin and is now used as an office building. Headquartered in Shanghai, Jardine, Matheson & Co. Ltd. was the first British company which sold opium to China. Its Tianjin branch was set up at Zizhulin Wharf in 1867. When it was first set up, the Tianjin Branch of Jardine, Matheson & Co, Ltd. was mainly a shipping agency. But in 1872, it reorganized the ships under its management and set up Huahai Shipping Company. When Jardine, Matheson Shipping Company was established in Shanghai in 1881, the shipping department in its Tianjin branch was changed into Tianjin branch of Jardine, Matheson Shipping Company accordingly. This two-storied building was built in 1921, covering an area of 1,800 square meters and having a floor area of about 2,800 square meters. The whole building is in a perfect symmetry with the entrance right in the middle of the facade and a Corinthian column on each side. Plain exterior design and elaborate interior decoration label the building as a Classical building.

外檐局部 Partial outer eaves

外景 Exterior view

内檐门 Gate under the inner eaves

老天花板 Old-fashioned ceiling

↓走廊 Corridor

内檐柱头 Column capital under the inner eaves

楼梯柱头细部 Detail of the Column capital of stairs

楼梯 Stairs

Former Location of Butterfield & Swire

原太古洋行大楼

原太古洋行大楼位于和平区解放北路165号，天津市区、县文物保护单位，重点保护等级历史风貌建筑，现为办公用房。太古洋行开设于1867年，由英商斯维尔与巴特费尔德联合创办，总行设在上海，曾与怡和洋行一起长期垄断中国航运业。天津分行于1881年开业，主要经营进出口贸易和远洋航运。该建筑建于1895年，占地面积约1 000 m^2，建筑面积约2 200 m^2。砖木结构二层楼房，房间宽敞，室内装饰简洁。外檐以青砖为主，窗套等部位用红砖相间，体现了中国传统建筑材料的丰富。二楼以壁柱装饰，用拱形门窗作为主要形象特征，并辅以放射状花饰丰富其细部。入口位于大楼中部，采用内凹手法，以突出建筑的壮观，入口门为旋转门，至今仍在使用。

历史照片 Historic photo

正立面渲染图 Facade Rendering

Situated at 165 North Jiefang Road, Heping District, the former location of Butterfield & Swire is the heritage site under city/county protection and the important protection historic architecture in Tianjin and is now used as an office building. Butterfield & Swire was opened jointly by British businessmen Swire and Buttrefield in Shanghai in 1867. It had monopolized Chinese shipping industry together with Jardine, Matheson & Co, Ltd. The Tianjin Branch of Butterfield & Swire was set up in 1881 and engaged in foreign trade and ocean shipping. This two-storied brick-wood building with spacious rooms was put up in 1895, covering a floor space of some 1,000 square meters and having a floor area of about 2,200 square meters. The exteriors are dry walls and red bricks are used on the window trims. The second floor features pilasters, arch windows and radial patterns. The building is symmetrically designed with the entrance in the middle, shaping like a Chinese character “凹”. The revolving door at the entrance is still in use.

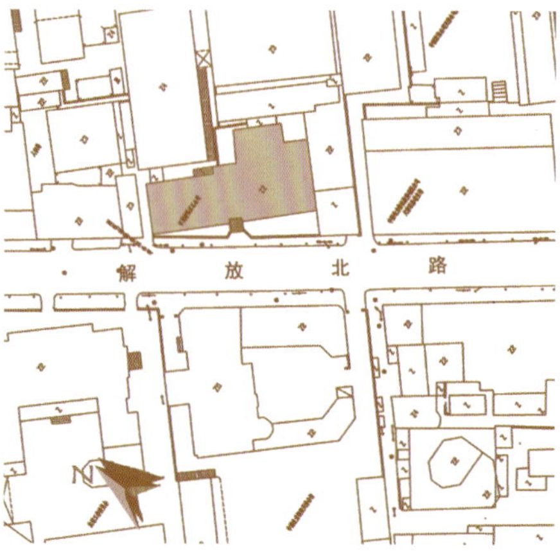

外景 Exterior view

外檐局部 Partial outer-eaves

外景 Exterior view

Former Rerun Firm

原瑞隆洋行

原瑞隆洋行位于和平区解放北路197号，重点保护等级历史风貌建筑，现为办公用房。瑞隆洋行开设于二十世纪二十年代，由英国人戴维斯创办。戴维斯是天津英商赛马会非正式骑师，受天津实业家高星桥资助开办瑞隆洋行，主要经营外汇、黄金、股票、房地产业务，为天津主要证券交易所之一。该建筑占地面积约800 m^2，建筑面积约1 700 m^2。砖木结构二层楼房，室内装饰简洁。外檐为红砖清水墙面，正立面对称，首层为混水饰面，方窗、拱券门。二层为拱券窗，窗间饰以简化爱奥尼克壁柱。建筑转角处均设罗汉腿壁柱装饰。坡屋顶，上出老虎窗，屋顶两侧各有一小型穹窿顶塔楼，使建筑更加活泼灵动。

外景 Exterior view

历史照片 Historic photo

Located at 197 North Jiefang Road, Heping District, Tianjin, the former Rerun Firm is now an important protection historic architecture in Tianjin and used as an office building. In the 1920s, Davis, an amateur horseman from the British Businessman of Horse Race Club, set up the firm in the aid of Gao Xingqiao, a local industrialist. It was one of the major stock exchanges and mainly engaged in foreign exchange, gold, stocks and real estate. This two-storied brick-wood structure covers an area of some 800 square meters and has a floor area of about 1,700 square meters. Red brick walls form the symmetric facade. The first floor features furred walls, square windows and arch gate while the second floor features Ionic pillars and arch windows. Pilasters are found at every corner of the building. The dormant windows and two vaulted towers standing on the pitched roof give the building a lively look.

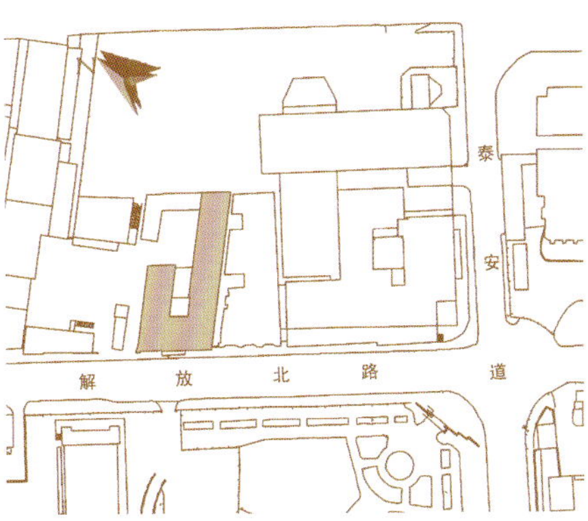

外景 Exterior view

Former Location of American-Chinese Company Federal Inc.

原美丰洋行

原美丰洋行位于和平区解放北路22~26号，重点保护等级历史风貌建筑，现为办公用房。美丰洋行有限公司为1922年成立的美商公司，总部设在华盛顿，1925年5月成立天津分公司，专营进出口汽车、汽车零配件、汽车修理等业务。该建筑建于1926年，占地面积约3 000m²，建筑面积约4 200m²。混合结构三层楼房，内部用于办公兼有居住功能，装饰简洁。沿街立面大部分为混水墙面，其余立面均为红砖清水墙面。沿街立面一层为骑楼，二、三楼顶部出牛腿支撑，装饰以精美花饰，三楼设阳台，设有宝瓶、铁艺两种栏杆。建筑设计尺度适宜，构图和谐，具有折中主义建筑特征。

外景 Exterior view

Located at 22~26 North Jiefang Road, Heping District, Tianjin, the former Location of American-Chinese Company Federal Inc. is an important protection historic architecture in Tianjin and it serves as an office building now. Established in 1922, the American-Chinese Company Federal Inc. was an American company headquartered in Washington DC. In May 1925, the company established its branch in Tianjin specializing in automobile import and export, auto spare parts and auto reparation, etc. Constructed in 1926, the building covers an area of 3,000 square meters and has a floor area of 4,200 meters. Succinctly decorated, it is a three-storied mixed structure suitable for both office and residence. Most parts of the building facing the street are painted walls and the rest are dry walls made of red bricks. The first floor facing the street is arcade while the top of the second and third floor are supported by corbel and finely decorated. On the third floor is a terrace equipped with iron banisters. The building was appropriately and harmoniously designed, showing the characteristics of Eclecticism.

外景 Exterior view

正立面渲染图 Facade Rendering

外檐局部 Partial outer-eaves

壁炉 Fireplace

内檐门 Gate under the inner eaves

↓楼梯 Stairs

↓楼梯间 Staircase

Entertainment Buildings

文娱建筑

Guangdong Guild Hall

广东会馆

广东会馆位于南开区南门里街31号，全国重点文物保护单位，特殊保护等级历史风貌建筑，现为戏剧博物馆。该会馆建于1907年，由天津海关道唐绍仪等倡议集资兴建，是旅津广东人议事集会的场所。该建筑群是天津现存规模最大、装饰最精美的会馆。会馆内的戏楼装饰精美，音响效果好。1912 年孙中山曾在这里发表演说；1919年邓颖超曾领导觉悟社成员在这里进行募捐义演；1925年中共领导的天津总工会在这里成立；孙菊仙、杨小楼、梅兰芳、荀慧生等京剧大师都曾在此馆演出。整个建筑群采用中国传统的合院建筑形式，具有广东潮州四合院的建筑特征；砖木结构，青砖饰面，重要的房间和门窗均有精美的砖雕、木雕，工艺精湛；会馆内的戏楼采用穹隆状密集旋转斗拱藻井，装饰华美。

砖雕 Brick carvings

The Guangdong Guild Hall is located at 31 Inner Nanmen Street, Nankai District, classified as the important heritage site under state protection and the special protection historic architecture in Tianjin, and it is used as theater museum right now. The Hall was constructed in 1907, sponsored by Tang Shaoyi, Governor of Tianjin Customs, and his fellows. It served the people from Guangdong Province for their gatherings. It is regarded as the biggest and best-decorated guild-building complex among this type of the constructions preserved in Tianjin. And it also had got its theater present with the most tasteful environment and acoustics. It was the place where Dr. Sun Yatsen gave his speech in 1912, Madam Deng Yingchao with the members of her Consciousness Organization came to send round the hat in 1919 and the Tianjin Federation of Trade Unions under the leadership of the Chinese Communist Party was founded. Moreover, many masters of the Peking Opera gave their performances here, such as Sun Juxian, Yang Xiaolou, Mei Lanfang and Xun Huisheng, etc. The complex is a design of a Chinese traditional enclosing layout, further with the characters of the courtyard houses in Chaozhou Area of Guangdong Province. It is brick-wood structured with the fair-faced walls. Elegant carvings made of brick and wood would be seen anywhere from the main rooms and their doors/windows. Brackets and coffers in special shapes were arranged densely against the arched roof, in order to enhance the effects not only of the decoration but also of the acoustics.

内院 Inner yard

↓内院回廊 Cloister

瑞兽 Auspicious-beast ornament

具有南方特色的马头墙 Horse-head-shaped wall in Southern style

外景 Exterior view

山墙砖雕 Brick carvings on the gable

外檐局部 Partial outer-eaves

戏楼内景 Interior view of the Theatre

戏台藻井 Coffered ceiling above the stage

↓戏台木雕 Wood carvings of the stage

精美的木雕 Artistic wood-carvings

壁画 Mural paintings

柱础 Column base

Confucian Temple
文庙

文庙也称孔庙，为祭祀孔子的场所。天津文庙位于南开区东门里大街1号，天津市文物保护单位，特殊保护等级历史风貌建筑。该建筑群由府庙、县庙、明伦堂组成，按照中国传统的古建筑营造体制建造。文庙始建于明正统元年（1436年），当时称卫学，清雍正年间，天津卫逐渐升为府，该庙则归天津府管理，称府庙。天津府的首县即天津县，因府、县官员不可同祭孔子，雍正十二年（1734年）在府庙西侧另建文庙，由天津县管理，称县庙。由此形成了天津文庙府、县并存的格局。两庙均由万仞宫墙、泮池、棂星门、大成门、大成殿、崇圣祠和东西两侧配殿等组成。县庙为青砖青瓦，体量较小。府庙体量较大，中轴线建筑为黄琉璃瓦顶。文庙是明清两代至民国初年，天津于春秋两季举行祭孔大典的场所，现为文庙博物馆。

大殿外景 Exterior view of the Grand Hall

鸟瞰图 Bird’s eye view

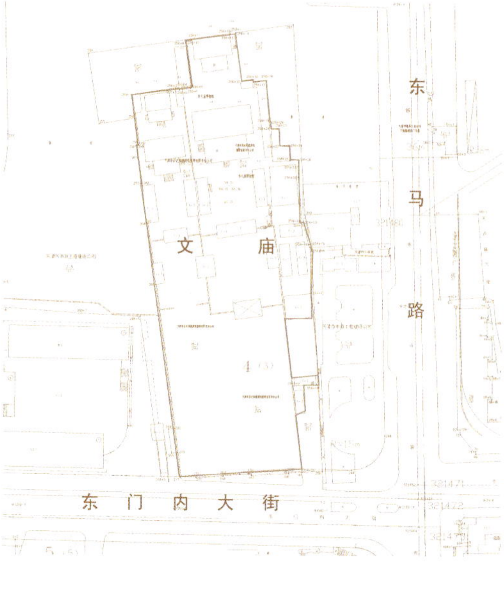

府庙大成门 Dacheng Gate of the Prefecture Temple

府庙棂星门 Lingxing Gate of the Prefecture Temple

府庙大成殿 Dacheng Hall of the Prefecture Temple

Confucian Temple (Wen Temple) is a place to offer sacrifices to Confucius. Standing at 1 Inner DongMen Street, Nankai District, Tianjin, the temple is the heritage site under city protection and the special protection historic architecture in Tianjin. The temple, constructed by following the style of traditional Chinese architectural complex, consists of Prefecture Temple, County Temple and Minglun Hall. Established in 1436 (the first year reign of Zhengtong Emperor in Ming Dynasty), the Confucian Temple was known as a garrison school (Wei Xue) at that time. During the reign of Yongzheng Emperor of Qing Dynasty, Tianjin was promoted from a garrison to a prefecture and the temple was put under the control of Tianjin prefecture. Since then it started to be called the Prefecture Temple (Fu Miao).

In ancient China, officials of different levels couldn't offer sacrifice to Confucius in a same temple and the first county under the Tianjin prefecture's control is Tianjin County. Therefore,

in 1734 (the twelfth year reign of Yongzheng Emperor of Qing Dynasty), the Confucian Temple (Wen Miao) was established to the west of the Prefecture Temple (Fu Miao) and was called County Temple (Xian Temple) under the control of Tianjin County. This is why two Confucian Temples at the county and prefecture level coexisted in Tianjin. Both of the two temples consist of the Main South Gate (Wanren Gongqiang), Pan Pond (Pan Chi), Lingxing Gate, Dacheng Gate, Dacheng Hall, Chongsheng Ancestral Hall and Eastern and Western Side Hall, etc. Decorated with grey bricks and tiles, the County Temple (Xian Temple) is smaller compared with the Prefecture Temple (Fu Temple) whose buildings along the axis of the complex are decorated with yellow glazed tiles. In Ming, Qing Dynasties and the early period of Republic of China, Confucian Temple (Wen Temple) was places to offer sacrifices to Confucius in spring and autumn. Now it is the Confucian Temple Museum.

县庙大成殿天花板 Ceiling of the Dacheng Hall of the County Temple

↓大殿内景 View inside the Grand Hall

县庙崇圣祠 Chongsheng Ancestral Hall of the County Temple

明伦堂 Minglun Hall

↓府庙西庑 Western Side Hall of the Prefecture Temple

县庙大成殿鸱吻 Chiwen of the Dacheng Hall of the County Temple

泮池 Pan Pond

下马石 Dismounted stone

县庙西侧礼门义路坊 Yilu Memorial Archway of Li Gate on the West of County Temple

府庙礼门义路坊 Yilu Memorial Archway of Li Gate of the Prefecture Temple

↓过街牌楼 Arcade

外景 Exterior view

Former French Club

原法国俱乐部

原法国俱乐部位于和平区解放北路29号，天津市区、县文物保护单位，重点保护等级历史风貌建筑，现为办公用房。法国俱乐部成立于十九世纪九十年代，是法国侨民娱乐的中心，也是法国商人议事聚会的场所，1932年法租界工部局兴建该楼，占地面积约7 000 m^2，建筑面积约2 900 m^2。该建筑为一层砖混结构，带有地下室。正门设于临街转角处，门两侧有逐段收分的竖向线条和附墙多边形壁灯柱，甚为别致。楼内设有角形中央大厅、舞厅、酒吧、球房、多功能厅等，功能齐备，可进行击剑、拳击、小型演出等活动。建筑外型简洁，门、窗户等元素尺度得当，铁艺精美，体现了十九世纪末新工艺艺术运动的特色和现代建筑的萌芽。

历史照片Historic photo

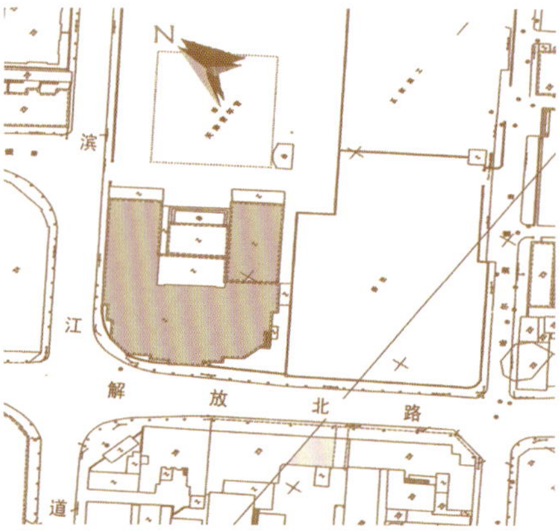

The Former French Club is situated at 29 North Jiefang Road, Heping District. It is now the heritage site under city/county protection and the important protection historic architecture in Tianjin and is used as an office building. The club was founded in the 1890s and became the place where the French people gathered to entertain and discuss over important issues. In 1932, the Industry Bureau in French Concession started constructing the building which occupies an area of some 7,000 square meters and has a floor area of about 2,900 square meters. The main entrance of this one-storied house is at the corner and it has a basement. With vertical lines and polygon lamp columns on both sides of the entrance, it looks very special. The building has an angle central hall, a dance hall, a bar and a multifunctional hall where people could practice fencing, boxing and other performances. The plain design, good-sized doors and windows and delicate wrought iron all demonstrate the characteristics of the art movement at the end of the ninetieth century and the emergence of modern architectures.

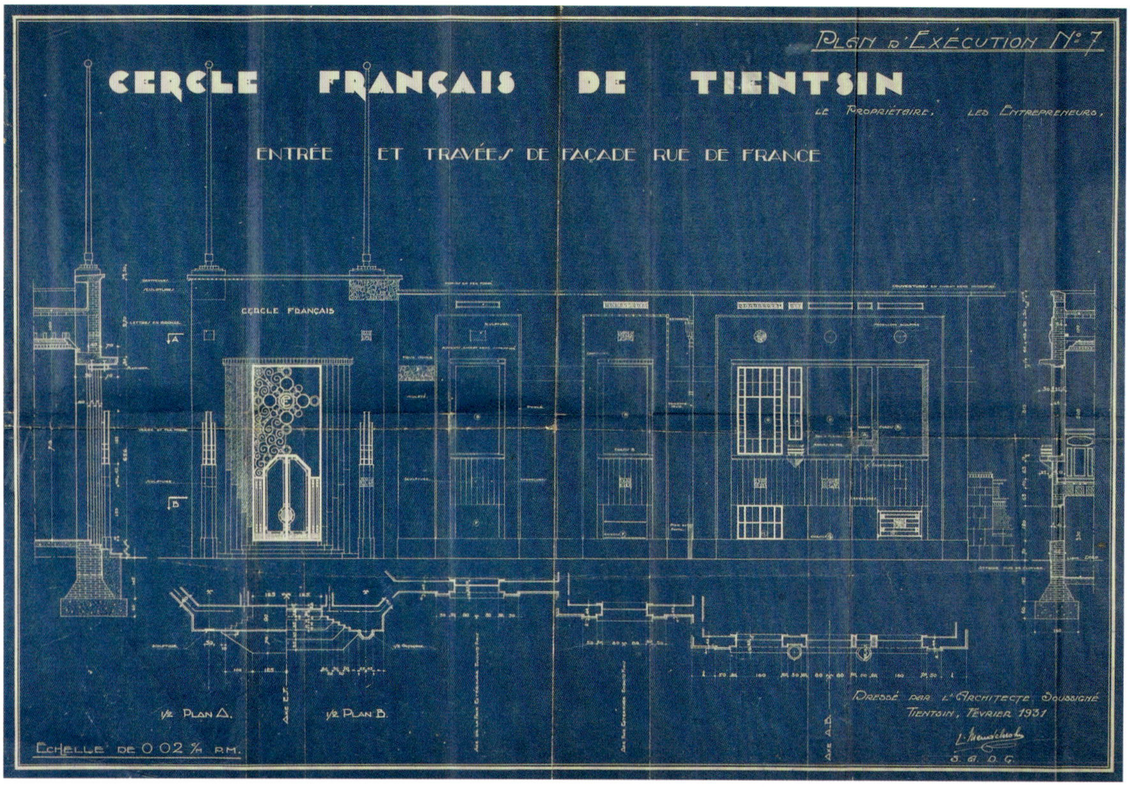

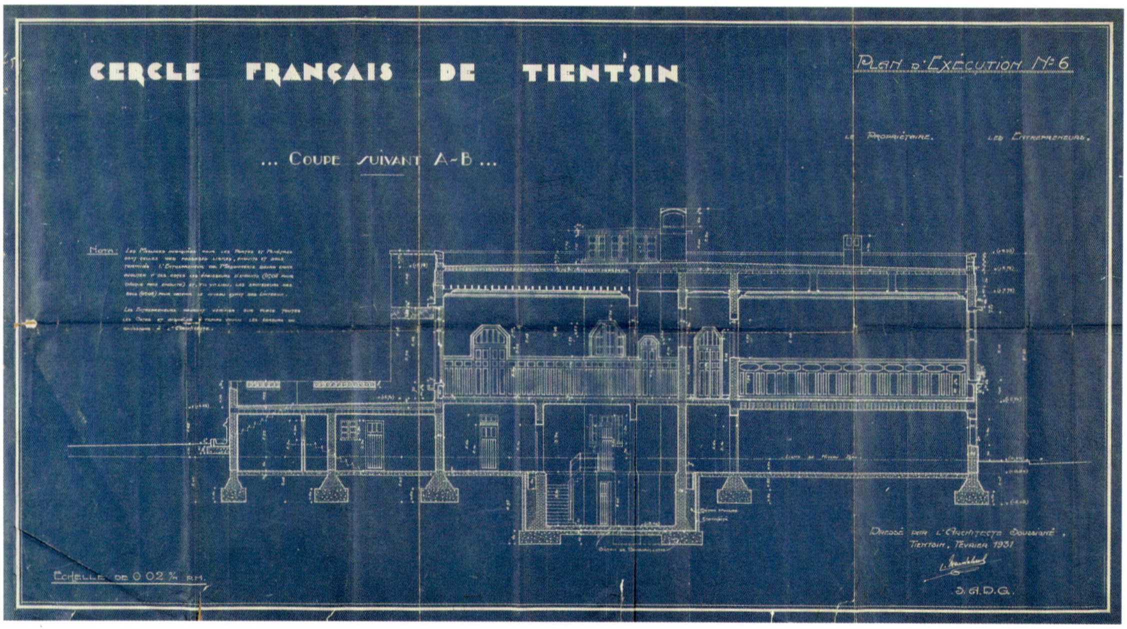

原设计图 Original blueprint

正立面渲染图Facade Rendering

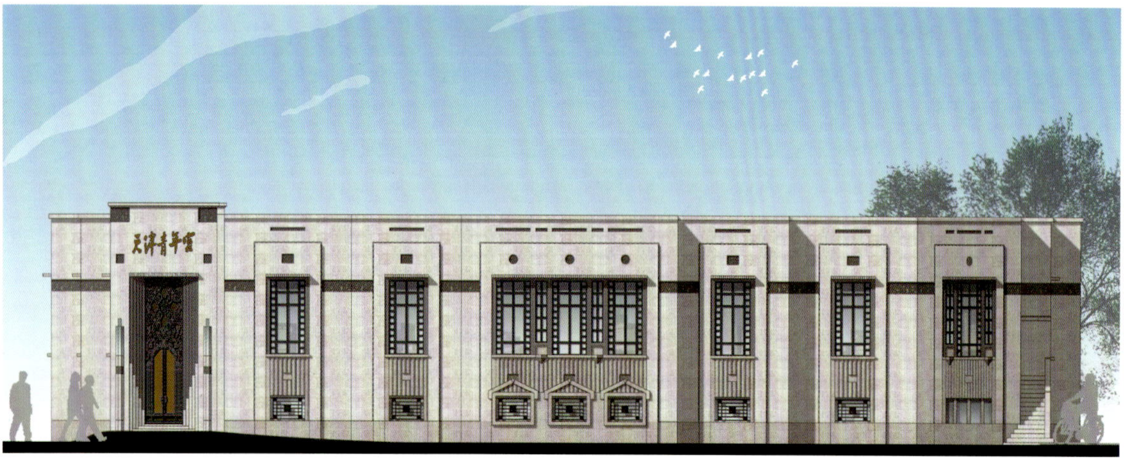

主入口 Main Entrance

一层大厅 lobby on first floor

Former Tientsin Club

原英国俱乐部

英国俱乐部初建于1860年前后，也叫英国球房，附设兰心戏院。系英租界侨民集资修建。每逢天津港封冻期间，侨民们即集中在俱乐部中举行舞会、午后茶会、音乐会或戏剧演出，是当时天津租界社会的艺术中心，现为办公用房。现存建筑为1904年重建，位于和平区解放北路201号，天津市文物保护单位，特殊保护等级历史风貌建筑，占地面积约7 300m^2，建筑面积约3 000m^2。砖木结构二层楼房（设有地下室）。建筑采用对称布局，窗间墙均匀布置的爱奥尼克巨柱强调了竖向构图，丰富的立面装饰和拱券门窗，体现了典型的折中主义建筑风格。

历史照片 Historic photo

The Club, also called the British Ballroom, was initially opened about 1860 accompanied with the Lyceum Theatre. Raised financially by the immigrants of the British Concession in Tianjin, the Club served the British immigrants as an entertainment center for holding balls and pink-tea parties, as well as enjoying concerts and operas.

The preserved building was reconstructed in 1904, located at 201 North Jiefang Road, Heping District and listed as the heritage site under city protection and the special protection historic architecture in Tianjin. The building is used as office building nowadays, covering an area of about 7,300 square meters, and about 3,000 square meters in floor area.

The brick-wood composite construction is two-storied, plus its basement. As a symmetrical design, the building attracts people's attention to its vertical structure by locating the massive Ionic columns orderly between the windows. Rich in exterior decoration and variation of the arched windows, it tastes to be typically in Eclectic style.

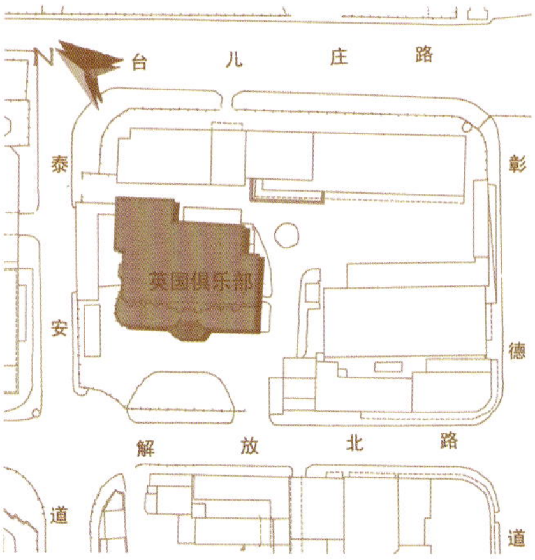

外檐局部 Partial outer-eaves

外檐门 Gate under the outer-eaves

↓外檐窗 Windows below the outer-eaves

外檐局部 Partial outer-eaves

外景 Exterior view

天津市人民代表大会常务委员会

一楼大厅 Lobby on first floor

↓主入口旋转门 Revolving door at the Main Entrance

↓内檐门 Gate under the inner eaves

二楼大厅顶棚 Ceiling of Lobby on the seond floor

一楼天花板 Ceiling of the first floor

↓老暖气 Old-fashioned radiator enclosure

Former Club Concordia
原德国俱乐部

原德国俱乐部位于河西区解放南路273号，天津市文物保护单位，特殊保护等级历史风貌建筑，现为办公用房。建筑建于1907年，又名德国会馆或康科迪亚俱乐部，是德国侨民政治、社交和文娱活动中心。二楼走廊墙壁刻有解释“康科迪亚”含义的拉丁文：“同心同德则盛，离心离德则衰。”该建筑占地面积约3 100m²，建筑面积约3 500m²，砖木结构三层坡顶楼房，楼内大厅和过道都以半圆券和椭圆形券承重，楼梯立柱和栏杆均饰以精美的雕刻，门窗造型多采用拱形元素。建筑装饰丰富，细部精致，具有明显的日耳曼建筑风格。

历史照片 Historic photo

The preserved site of the Club Concordia was located at 273 South Jiefang Road, Hexi District, classified in the list of the heritage site under city protection and the special protection historic architecture in Tianjin, covering an area of about 3,100 square meters, and about 3,500 square meters in floor area.

The architecture was built in 1907. The Club Concordia, also known as the German Club, served mainly for the German immigrants as their political, social and recreational center. Some words in Latin were inscribed on the wall of second floor, explaining the mean of the Club: if “Concordia”—“all the people join their efforts for common purpose” will bring out prosperity; and if “discordia”-everyone only concerns oneself might lead to collapse.

The brick-wood three-storied building is designed with the sloping roofs. Structure in arch shape could be seen anywhere, such as the arched doors, arched windows and arched roofs. Elegant carvings were adapted to the stair poles and balusters. It tastes typically to be the architecture in Nordic style.

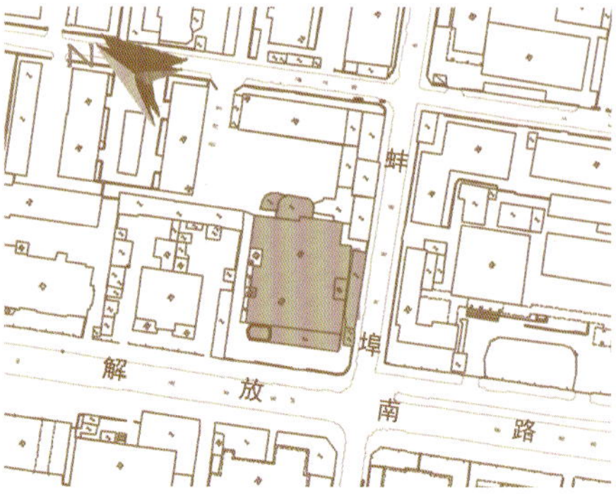

外景 Exterior view

内景 Inside view

大厅楼梯 Stairs in the Lobby

一楼餐厅 Restaurant at first floor

拱券顶走廊 Corridor with the arched ceiling

一楼餐厅 Restaurant at first floor

Former American Navy Club

原美国海军俱乐部

原美国海军俱乐部位于解放北路113号，天津市区、县文物保护单位，重点保护等级历史风貌建筑，现为办公用房。该建筑建于1924年，占地面积约500 m^2，建筑面积约1 400 m^2。为英国人罗士博士开办的专供外国驻军娱乐的场所。美国海军于1945年在此设俱乐部，楼内设有酒吧、球房、餐厅、赌场、咖啡厅、舞厅等。建筑外檐装饰丰富，入口处有醒目凸出的阳台，拱券窗、扶壁柱等建筑元素交相辉映，具有古典主义建筑特征。

外景 Exterior view

历史照片 Historic photo

外檐窗套细部 Details of the window trim under the outer-eaves

Located at 113 North Jiefang Road, Heping District, the former American Navy Club is now the heritage site under city/county protection and the important protection historic architecture in Tianjin and used as an office building. Constructed in 1924 and covering an area of some 500 square meters and having a floor area of about 1,400 square meters, this building was owned by Dr. Ross and served as an entertainment place for foreign soldiers. In 1945, this building was used by American Navy as a club and equipped with various services such as bar, billiard rooms, restaurants, casino, coffee house and dancing hall. The building has beautifully decorated roofs, an eye-catching balcony over the entrance, arch-windows and buttress, and is a typical classical building.

Former Tianjin Race Club

原英国乡谊俱乐部

原英国乡谊俱乐部位于河西区马场道188号，天津市文物保护单位，特殊保护等级历史风貌建筑，现为商业用房。建筑建于1925年，建筑面积约8 000m^2，英商景明工程司设计，又称新英国俱乐部。俱乐部包括剧场、舞厅、室内游泳池、乒乓球室、保龄球室、图书馆等多项游乐设施，是外国侨民和上流社会娱乐消遣的场所。1941年太平洋战争爆发后被日军占用，改名“国际俱乐部”。1945年日本投降后改为美军“军官俱乐部”。该建筑为混合结构庭院式二层楼房，建筑立面呈对称形式，造型稳重，红砖饰面朴素大方。内部装饰精美适用，各项娱乐功能在当时的天津首屈一指。乡谊俱乐部坐落于英国跑马场内，周围园林景观优美，建筑具有浓郁的英国田园风格。

外景 Exterior view

Standing at 188 Machang Avenue, Hexi District, the former Tianjin Race Club is now the heritage site under city protection and the special protection historic architecture in Tianjin and is used as a business building. It was built in 1925 and has a floor area of some 8,000 square meters. As was designed by British Hemmings & Parkin Ltd., it was also known as New British Club. Equipped with theatre, dance hall, indoor swimming pool, pingpang room, bowling alley and library, the club was exclusive to foreigners and the upper crust. After the Pacific War broke out in 1941, the Japanese invaders took it over and renamed it as International Club. And after Japan's surrender in 1945, it fell into American's hand and was changed into Armed Forces Officer's Club. Featured a simple red brick facade with a perfect symmetry and beautiful interior decorations, the facilities in this two-storied building topped its kind in Tianjin at that time. Located inside the race ground in the British concession featured a beautiful garden landscape, the building epitomized British pastoral style.

历史照片 Historic photo

楼梯 Stairs

↓一楼大厅 Lobby on first floor

木雕细部 Detailed wood carving

吊灯 Hanging lamp

老穿衣镜 Old-fashioned mirror

壁灯 Wall lamp

内檐入口门 Entrance under the inner-eaves

壁灯 Wall lamp

舞厅回廊 Cloister of the dance hall

↓老游泳池 Old swimming pool

老保龄球道
Old-fashioned bowling lanes

老台球厅记分牌
Old billiard scoreboard

Great China Theatre

中国大戏院

中国大戏院位于和平区哈尔滨道104号，天津市文物保护单位，重点保护等级历史风貌建筑。该戏院由周振东（法租界工部局翻译）与天津商人孟少臣投资修建，土地由著名外交家顾维钧捐赠。1936年9月建成开业时，时为市长的张自忠将军参加了揭幕典礼。由马连良先生领衔的扶风社做了首场演出。梅兰芳、尚小云、周信芳等著名艺术家陆续出演，盛况空前。该建筑由乐利工程司设计，占地面积约2 700 m^2，建筑面积约7 700 m^2，是当时天津规模最大、设备最新的戏院。该建筑为混合结构四层楼房，局部五层，外檐为混水墙面，立面采用多组垂直竖线，使建筑呈现挺拔向上的效果，具有现代建筑简洁明快的风格。内部装饰考究，布局合理，剧场的建筑声学效果极佳。

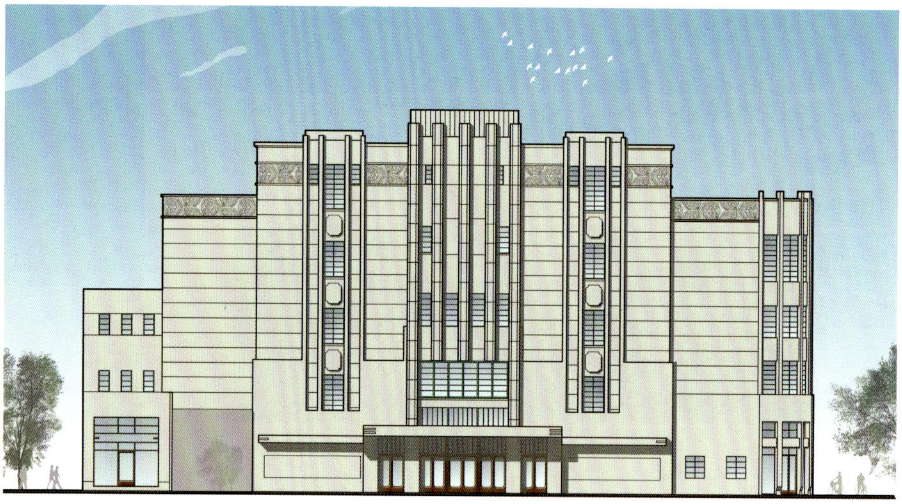

正立面渲染图 Facade Rendering

Situated at 104 Haerbin Avenue, Heping District, the Great China Theatre is the heritage site under city protection and the important protection historic architecture in Tianjin. Zhou Zhendong (The translator of French Industry Bureau) and Meng Shaochen (A local businessman) invested to build the theatre on the land donated by Gu Weijun, an eminent diplomat. In Sep. 1936, General Zhang Zizhong, the then mayor, attended the opening ceremony and the Fufeng Troupe led by Ma Lianliang was invited to give the first performance in the history of the theatre. Covering an area of some 2,700 square meters and having a floor area of about 7,700 square meters, the theatre designed by Leli Engineering Company was the largest and most advanced theatre in Tianjin at that time. The vertical lines on the facade make the four-storied building look straight and high, reflecting the simple style of modern architecture. Moreover, the theatre features a top-class architectural acoustics design.

外景 Exterior view

外景 Exterior view

入口大厅 Lobby at the entrance

楼梯 Stairs

↓楼梯扶手 Stair rail

戏院内景 Inner view of the theatre

老电梯 Old-fashioned elevator

老电梯内景 Inner view of the Old-fashioned elevator

老电梯铭牌 Nameplate of the old-fashioned elevator

老电梯指针 Indicated needle of the old-fashioned elevator

老电梯按钮 Button of the old-fashioned elevator

展厅 Lobby

夜景 Night view

Former Jaialai Sportcenter

原回力球场

原回力球场位于河北区民族路47号，重点保护等级历史风貌建筑，现为商业用房。回力球场由意商富马加里创办，是借回力球运动赌博的场所，内设赛场、餐厅、休息室等，是当时华北地区最大的室内游乐场。1942年，球场改由瑞士人李亚溥经营，改名为海莱运动场，抗日战争胜利后停业。该建筑建于1934年，占地面积约4 000 m^2，建筑面积约6 400 m^2，意大利建筑师鲍乃弟和瑞士人凯思乐设计，孟特劳克公司施工。新中国成立后，由关颂坚和凯思乐重新设计改建为工人文化宫。建筑为钢混结构楼房，立面设计强调竖向构图，具有现代建筑特征。角部塔楼在1976年地震中损毁，2008年恢复。

The Former Jaialai Sportcenter, situated at 47 Minzu Road, Hebei District, is now an important protection historic architecture in Tianjin and is used for business. It was a gambling house started by V. Fumagalli and was the largest indoor sport center with sports ground, dining hall and lounge in North China. In 1942, M.LEOPOLD, a Swiss took over the center and renamed it as Hailai Sportcenter. After the victory in 1945, it was closed down. The center was designed by an Italian architect Bonetti and a Swiss called Kessler and was built in 1934, covering an area of some 4,000 square meters and having a floor area of about 6,400 square meters. After the new China was founded, it was redesigned by Guan Songjian and Kessler and rebuilt into the Workers' Club. This steel-concrete structure building is very modern because the vertical format on the facade. In the 1976 Tangshan Earthquake, the tower on the roof was damaged and was restored in 2008.

夜景 Night view

历史照片 Historic photo

外景 Exterior view

外景 Exterior view

Former Japanese Pavilion

原武德殿

原武德殿位于和平区南京路228号，天津市文物保护单位，特殊保护等级历史风貌建筑，现为办公用房。武德殿又名演武馆，始建于1941年，建筑面积约2 200m^2，是日本武德会天津支部为日本驻军和日本侨民所建习武健身的场所。首层为公寓式住房，二楼为练武厅，内设柔道、击剑、拳击等健身习武设施。该建筑为二层砖木结构日本和式建筑，坡屋顶瓦屋面，首层外墙粘贴釉面砖，二层为混水墙面。建筑体量较大，典雅大方，稳重舒展，强调传统建筑的对称构图。经过长期使用并经历了唐山大地震，该建筑的屋顶构件出现了损坏和缺失，在2008年迎奥运环境综合整治工作中，对该建筑进行了恢复原貌的整修，按原工艺重新定制了大量屋面构件并进行了更换修复，达到了“修旧如故”的效果。

Located at 228 Nanjing Road, Heping District, Tianjin, the former Japanese Pavilion is the heritage site under city protection and the special protection historic architecture in Tianjin. It is an office building now. Constructed in 1941, the pavilion, also known as Yanwu Guan, has a floor area of 2,200 square meters. The pavilion, established by the Tianjin branch of the Japan Judo Moral Association, was a place for Japanese soldiers and Japanese employees working in China to practice martial arts. The first floor of the building has some apartment-style rooms and the second floor serves as a martial arts practice hall where fencing, Judo and boxing facilities can be found. This building is a Japanese-style two-storied brick-wood structure with sloping roof covered with tiles. The exterior walls on the first floor are decorated with glazed bricks and the walls on the second floor are plastered brick walls. The building is symmetrical, elegant and spacious. Some components of the building's roof were damaged and lost due to the wear and tear of time and during the Tangshan earthquake. In 2008, the building was renovated thanks to the efforts made to improve the city's environment in preparation for the Olympic Games. Large numbers of roof components were replaced and installed to restore the building to its former state.

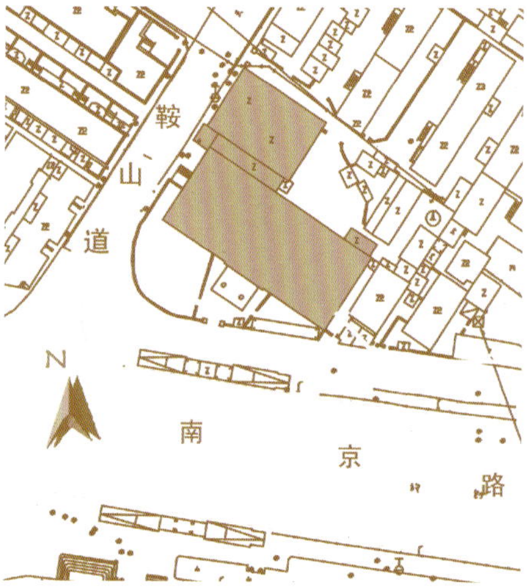

历史照片 Historic photo

外景 Exterior view

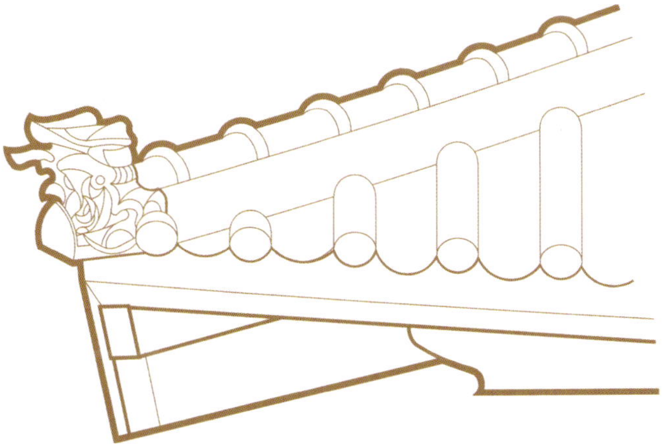

外景 Exterior view

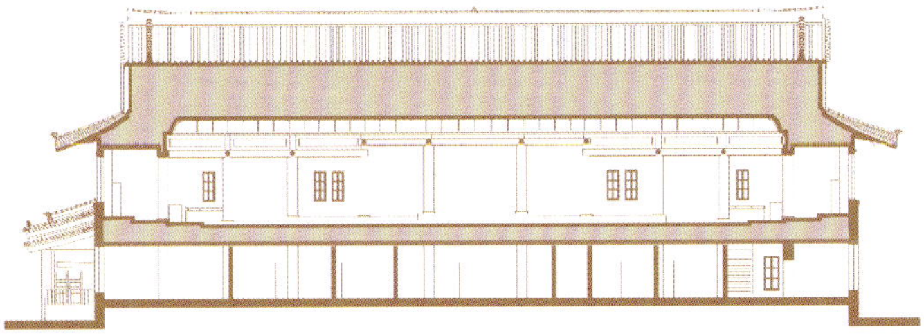

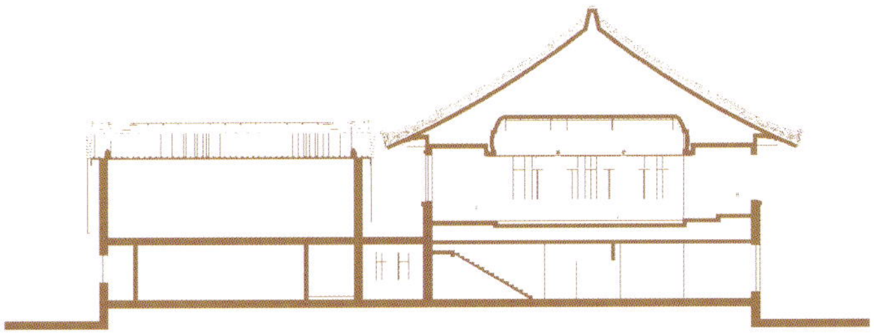

外檐局部 Partial outer-eaves

主入口 Main Entrance

Former Beijiang Museum

原北疆博物院

原北疆博物院坐落于河西区马场道117~119号，现天津外国语大学校内，天津市文物保护单位，重点保护等级历史风貌建筑，现为天津自然博物馆使用。该博物院是我国早期的博物院之一，由法国天主教耶稣教会桑志华神父于1914年创建，主要收藏和研究黄河、白河（海河）流域的地质、动植物，院名由此译为“北疆”。博物院由博物馆、试验馆两个建筑组成，建筑面积约2 000 m²，博物馆（北楼）于1922年由比商义品公司设计、建造，试验馆（南楼）于1925—1929年由法商永和营造公司承建，两楼间用封闭式天桥相连。建筑为混合结构三层楼房（局部二层），清水砖墙，立面简洁大方，平面布局满足功能要求。

历史照片 Historic photo

Located at 117~119 Machang Avenue, Hexi District, Tianjin, the former Beijiang Museum is now in Tianjin Foreign Language University and is the heritage site under city protection and the important protection historic architecture in Tianjin. Now it is used by Tianjin Natural Museum. Founded by the French priest Emile Licent in 1914, the museum is one of the first museums established in China. The museum was called Beijiang Museum because its collections were mainly geological, flora and fauna articles found along the Yellow River and Haihe River. Covering an area of 2,000 square meters, it consists of two parts, the Museum and the laboratory. The museum part (northern part) was designed and constructed by the Credit Foncier D'Extreme Orient in 1922. The laboratory part (southern part) was constructed by French Yonghe Company between 1925 and 1929. The two parts are connected by an enclosed overpass. The building is a three-storied (part of it is two-storied) mixed structure with fair-faced walls. The facade of the building is simple and natural. The layout meets the required function of the building.

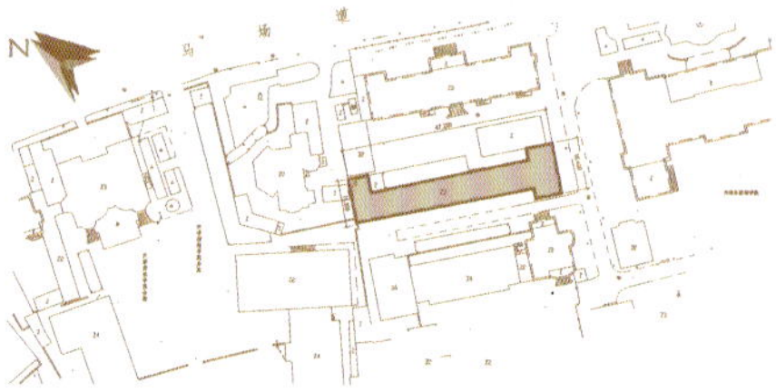

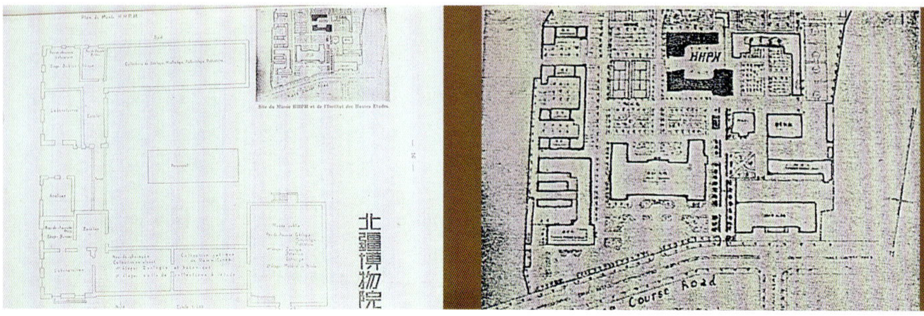

北疆博物院和原工商学院老平面图
Historic plan of the Musee Hoang-Ho Pai-Ho and the Hautes Etudes Industrielles et Commerciales

现存北疆博物院老标牌
The historic tablet of the Musee Hoang-Ho Pai-Ho

外景 Exterior view

正立面渲染图 Facade Rendering

内檐门 Gate under the inner eaves

一楼仓库内景 Insight the Storage

馆内老水泥地面 Cement ground in the Museum

老锁 Old lock

Minyuan Stadium
民园体育场

民园体育场位于和平区五大道地区中部，由河北路、重庆道、衡阳路、大理道围合而成。该处原为洼地，1902年划入英租界。英租界工部局将疏通海河挖出的淤泥通过管道吹到这里，将洼地填平。1926年工部局将空地用铁栅栏围合，形成体育场。附近居民常来此进行体育活动，每年春秋季常有一些学校在这里召开运动会。万国足球赛和万国田径赛均曾在这里举行。1929年，天津出生的奥运会冠军—苏格兰人埃里克·利迪尔（中文名为李爱锐），在这里举办的“万国田径赛”上夺得了男子800米跑的金牌。

经过1954年、1974年、1980年至1982年几次大型的改建，修建了看台，台下设置了办公用房，场内铺设了草坪，民园体育场一度成为天津市设备最好的灯光球场。

Located at the center of the Fifth Avenue Area, Heping District, Minyuan Stadium is enclosed by Hebei Road, Chongqing Avenue, Hengyang Road and Dali Avenue. It was originally a low-lying piece of land, and was cut into the British Concession in 1902. The Municipal Council of British Concession conveyed the silt dredged from the Haihe River through tunnels to the location to fill and level up the land. In 1926, the Municipal Council enclosed the area with iron fences and formed the stadium. People living nearby often came for sports activities. Each year in spring and autumn, some schools held sports meet here. World Football Match and World Track and Field Meet were once held here. In 1929, the Olympic champion, Eric Henry Liddel (Chinese name Li Airui), born in Tianjin, won the gold medal of Men's 800 meters at the World Track and Field Meet held here.

After the founding of the People's Republic of China, in the large-scale reconstructions conducted in 1954, 1974, 1980 and 1982 respectively, grandstand was built; office rooms were set up under the stand; turf was laid in the field. Minyuan Stadium became the best illuminated stadium with advanced facilities for a time.

英租界当局为防止日本飞机轰炸租界，在民园体育场门前铺设国旗
British people laid a national flag on the ground in case of bombing

外檐背立面 Back elevation

↓主入口 Main entrance

鸟瞰图 Bird's eye view

Communication Buildings
交通建筑

Main House of the West Railway Station in Tianjin

天津西客站主站房

天津西客站主站房位于红桥区西站前街1号，天津市文物保护单位，特殊保护等级历史风貌建筑。天津西站建于1910年，是津浦铁路的重要站点。清光绪三十四年（1908年）清政府开始修建（天）津浦（口）铁路，1912年全线通车。该建筑建筑面积约1 700m^2，砖混结构二层楼房，建筑坐北朝南，平面呈凸字形，红瓦坡顶，清水砖墙，建筑立面强调对称构图，造型丰富，窗式、窗套富于变化，属典型的折中主义建筑风格。

历史照片 Historic photo

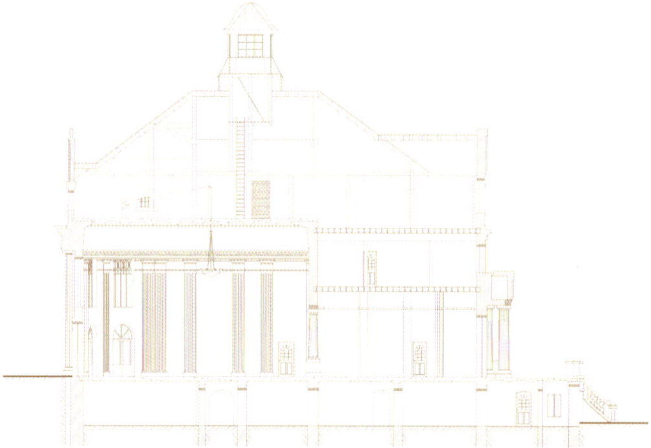

As the heritage site under city protection and the special protection historic architecture in Tianjin, the West Railway Station sits at 1 Front Street of the West Railway Station, Hongqiao District, and about 1,700 square meters in floor area.

The Station was built in 1910, serving as an important joint of the Jin–Pu (Tianjin–Pukou) railway which had initially been constructed in 1908(34th year of the Guangxu Reign, Qing Dynasty) and completed in 1912.

Facing southward, the two-storied brick-concrete construction has its facade look convex-shaped. It is in a symmetrical design with red-tiled sloping roof and dry brick walls, rich in structures, and windows along with their trims. The design presents typically in Eclectic style.

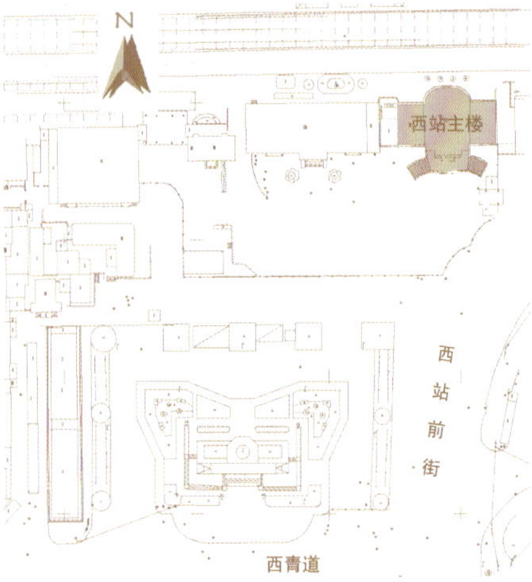

正立面渲染图 Facade Rendering

背立面局部 Partial of Back elevation

外檐局部 Partial outer-eaves

石雕栏板 Carved stone railing panel

外檐局部 Partial outer-eaves

外景 Exterior view

软席候车室
SOFT SEAT WAITING ROOM

候车厅 Lounge of the Station

Factories and Warehouses Buildings

厂房仓库建筑

Warehouse of the Former Continental Bank

原大陆银行仓库

二十世纪二十年代，大陆银行为使码头仓储和抵押贷款有所保证，修建了大型仓库，存放各种商货，从而夺得了外商70%的仓储业务。该建筑位于和平区张自忠路223号，特殊保护等级历史风貌建筑，建于1925年，建筑面积约7 700m^2。为四层（局部六层）混合结构楼房,清水砖墙，建筑体量厚重敦实，舒展大方，建筑内部设有壁垒森严的金库，现仍在使用。

A giant warehouse was set up by the Continental Bank in 1920s to further meet its operational requirements for storage and mortgage loan. 70% storage business of the foreign businessmen was obtained by the Bank therefore. Located at 223 Zhang Zizhong Road, Heping District, the warehouse built in 1925 has been classified as the special protection historic architecture in Tianjin. It covers a floor area of about 7,700 square meters. With dry brick wall the composite structure is mainly four-storied but partially six-storied. The building looks massive and decent. It was also designed with a well-guarded treasury which is still in use nowadays.

雪景 Snow view

外景 Exterior view

历史照片 Historic photo

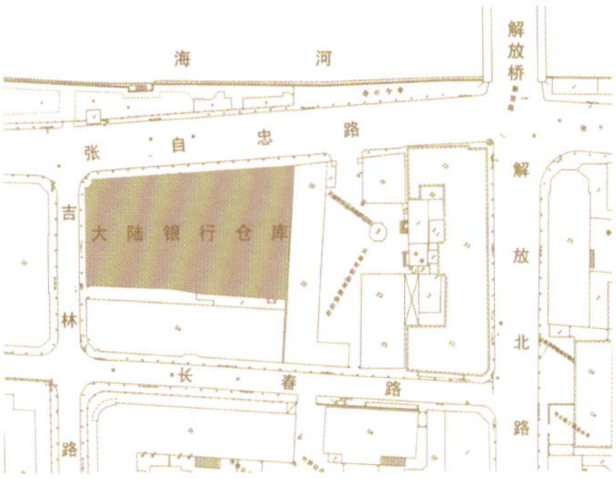

Former Site of Xigu Machine Factory of Jinpu Railway

津浦路西沽机场旧址

津浦路西沽机场旧址位于河北区南口东路1号，重点保护等级历史风貌建筑。津浦路西沽机场由德国人于1909年承建，是当时中国火车机车车辆修理的重要基地。现存厂房为两坡顶大罩棚机修车间，建筑面积约600m²，砖木结构，外檐墙为硫缸砖清水墙。每个开间由砖砌半圆拱券支撑，并由三联拱券组成窗口。砖砌竖向线条装饰墙面，四个转角处由砖砌成出屋顶砖柱。

The Former Site of Xigu Machine Factory of Jinpu Railway is located at 1 East Nankou Road, Heibei District, and listed as important protection historic architecture in Tianjin. Xigu Airport of Jinpu Railway was built by Germans in 1909, and became an important base of locomotives and carriages maintenance and repair base in the then China. The remaining workshop is a machine repair shop with a double-pitched huge awning, a brick-wood structure, a wooden roof truss and dry walls of glazed quarry bricks. It covers a floor area of 600 square meters. Each bay is supported by a semicircle brick arch while the window is composed of three joining arches. The brick-built vertical lines are adopted to decorate the walls. At the four corners are poles of roof made of bricks.

外景 Exterior view

图书在版编目（CIP）数据

公共建筑.2/路红主编. - 天津：天津大学出版社，2010.5
（天津历史风貌建筑）
ISBN 978-7-5618-3472-5
Ⅰ.①公… Ⅱ.①路… Ⅲ.①公共建筑：古建筑 - 建筑艺术 - 天津市 - 图集 Ⅳ.①TU - 092.2
中国版本图书馆CTP数据核字(2010)第071227号

组稿编辑　韩振平
责任编辑　韩振平

出版发行　天津大学出版社
地　　址　天津市卫津路 92 号天津大学内(邮编:300072)
电　　话　发行部:022-27403647　邮购部:022-27402742
网　　址　www.tjup.com
印　　刷　北京华联印刷有限公司
经　　销　全国各地新华书店
开　　本　210mm × 285mm
印　　张　17
字　　数　250千
版　　次　2010年5月第1版
印　　次　2022年1月第1版第2次印刷
定　　价　158.00元